The National Children's Study Research Plan

A REVIEW

Panel to Review the National Children's Study Research Plan

Committee on National Statistics
Division of Behavioral and Social Sciences and Education
NATIONAL RESEARCH COUNCIL

Board on Children, Youth, and Families
Division of Behavioral and Social Sciences and Education
NATIONAL RESEARCH COUNCIL AND INSTITUTE OF MEDICINE

and

Board on Population Health and Public Health Practice
INSTITUTE OF MEDICINE

NATIONAL RESEARCH COUNCIL *AND*
INSTITUTE OF MEDICINE
OF THE NATIONAL ACADEMIES

THE NATIONAL ACADEMIES PRESS
Washington, D.C.
www.nap.edu

NATIONAL ACADEMIES PRESS 500 Fifth Street, N.W. Washington, DC 20001

NOTICE: The project that is the subject of this report was approved by the Governing Board of the National Research Council, whose members are drawn from the councils of the National Academy of Sciences, the National Academy of Engineering, and the Institute of Medicine. The members of the committee responsible for the report were chosen for their special competences and with regard for appropriate balance.

This study was supported by Contract No. NO1-OD-4-2139 between the National Academy of Sciences and the National Institute of Child Health and Human Development. Support of the work of the Committee on National Statistics is provided by a consortium of federal agencies through a grant from the National Science Foundation (Number SBR-0453930). Any opinions, findings, conclusions, or recommendations expressed in this publication are those of the author(s) and do not necessarily reflect the views of the organizations or agencies that provided support for the project.

International Standard Book Number-13: 978-0-309-12056-2
International Standard Book Number-10: 0-309-12056-X

Additional copies of this report are available from National Academies Press, 500 Fifth Street, N.W., Lockbox 285, Washington, DC 20055; (800) 624-6242 or (202) 334-3313 (in the Washington metropolitan area); Internet, http://www.nap.edu.

Printed in the United States of America

Suggested citation: National Research Council and Institute of Medicine. (2008). *The National Children's Study Research Plan: A Review.* Panel to Review the National Children's Study Research Plan. Committee on National Statistics, Division of Behavioral and Social Sciences and Education. Board on Children, Youth, and Families and Board on Population Health and Public Health Practice, Institute of Medicine. Washington, DC: The National Academies Press.

THE NATIONAL ACADEMIES

Advisers to the Nation on Science, Engineering, and Medicine

The **National Academy of Sciences** is a private, nonprofit, self-perpetuating society of distinguished scholars engaged in scientific and engineering research, dedicated to the furtherance of science and technology and to their use for the general welfare. Upon the authority of the charter granted to it by the Congress in 1863, the Academy has a mandate that requires it to advise the federal government on scientific and technical matters. Dr. Ralph J. Cicerone is president of the National Academy of Sciences.

The **National Academy of Engineering** was established in 1964, under the charter of the National Academy of Sciences, as a parallel organization of outstanding engineers. It is autonomous in its administration and in the selection of its members, sharing with the National Academy of Sciences the responsibility for advising the federal government. The National Academy of Engineering also sponsors engineering programs aimed at meeting national needs, encourages education and research, and recognizes the superior achievements of engineers. Dr. Charles M. Vest is president of the National Academy of Engineering.

The **Institute of Medicine** was established in 1970 by the National Academy of Sciences to secure the services of eminent members of appropriate professions in the examination of policy matters pertaining to the health of the public. The Institute acts under the responsibility given to the National Academy of Sciences by its congressional charter to be an adviser to the federal government and, upon its own initiative, to identify issues of medical care, research, and education. Dr. Harvey V. Fineberg is president of the Institute of Medicine.

The **National Research Council** was organized by the National Academy of Sciences in 1916 to associate the broad community of science and technology with the Academy's purposes of furthering knowledge and advising the federal government. Functioning in accordance with general policies determined by the Academy, the Council has become the principal operating agency of both the National Academy of Sciences and the National Academy of Engineering in providing services to the government, the public, and the scientific and engineering communities. The Council is administered jointly by both Academies and the Institute of Medicine. Dr. Ralph J. Cicerone and Dr. Charles M. Vest are chair and vice chair, respectively, of the National Research Council.

www.national-academies.org

PANEL TO REVIEW THE NATIONAL CHILDREN'S STUDY RESEARCH PLAN

SAMUEL H. PRESTON (*Chair*), Department of Sociology, University of Pennsylvania
ELLEN WRIGHT CLAYTON, Center for Biomedical Ethics and Society and School of Law, Vanderbilt University
GREG DUNCAN, School of Education and Social Policy and Institute for Policy Research, Northwestern University
DAVID HARRINGTON, Department of Biostatistics, Harvard School of Public Health, and Department of Biostatistics and Computational Biology, Dana Farber Cancer Institute
RUSS HAUSER, Departments of Environmental Health and Epidemiology, Harvard School of Public Health
WILLIAM KALSBEEK, Department of Biostatistics, University of North Carolina
SHARON LEE REILLY KARDIA, Public Health Genetic Program, School of Public Health, University of Michigan
DANIEL KASPRZYK, Mathematica Policy Research, Inc., Washington, DC
MILTON KOTELCHUCK, Maternal and Child Health Department, School of Public Health, and Medical School, Boston University
MARIE McCORMICK, Department of Society, Human Development, and Health, Harvard School of Public Health
NORA S. NEWCOMBE, Department of Psychology, Temple University
PATRICIA O'CAMPO, Centre for Research on Inner City Health, St. Michael's Hospital, University of Toronto

GOOLOO S. WUNDERLICH, *Study Director*
BRIDGET EDMONDS, *Administrative Assistant*

BOARD ON CHILDREN, YOUTH, AND FAMILIES

BOARD ON POPULATION HEALTH AND PUBLIC HEALTH PRACTICE

Acknowledgments

The Panel to Review the National Children's Study (NCS) Research Plan acknowledges with appreciation the contributions of the many individuals who gave generously of their time and knowledge to this report.

Support for the review was provided by the Eunice Kennedy Shriver National Institute of Child Health and Human Development (NICHD). We particularly wish to thank Peter Scheidt, director of NCS, NICHD, who served as project officer for the review. He was very helpful in providing information and in responding to many questions from the panel and staff. We are grateful to Carla Maffeo, Westat, project officer for the NCS Coordinating Center, for speaking to the panel and providing relevant documents.

Members of the panel gave very generously of their time and expert knowledge in the deliberations that produced this report. Their timely and thoughtful work in drafting written reviews of the National Children's Study is gratefully acknowledged.

Staff members of the Committee on National Statistics (CNSTAT) and the Division of Behavioral and Social Sciences and Education (DBASSE) provided important support and assistance to the panel. We acknowledge with gratitude their contributions. Gooloo S. Wunderlich, study director, was responsible for organizing the meetings of the panel, preparing drafts of large portions of the report based on the reviews prepared by panel members, and responding to the many comments from the reviewers on behalf of the panel. Bridget Edmonds provided administrative support. We gratefully acknowledge the important role of Constance F. Citro, director of CNSTAT, in the preparation of this report. Christine McShane, senior editor of the DBASSE reports office, provided professional editing advice, and Kirsten Sampson Snyder efficiently shepherded the report through the report review process.

This report has been reviewed in draft form by individuals chosen for their diverse perspectives and technical expertise, in accordance with procedures approved by the Report Review Committee of the National Research Council. The purpose of this independent review is to provide candid and critical comments that will assist the institution in making its published report as sound as possible and to ensure that the report meets institutional standards for objectivity, evidence, and responsiveness to the study charge. The review comments and draft manuscript remain confidential to protect the integrity of the deliberative process.

We thank the following individuals for their review of this report: Lia C. H. Fernald, School of Public Health, University of California, Berkeley; Elizabeth T. Gershoff, School of Social Work, University of Michigan; Howard Hu, Department of Environmental Health Sciences, University of Michigan Schools of Public Health and Medicine; Aletha C. Huston, Human Development and Family Sciences, University of Texas at Austin; Lorraine V. Klerman, Institute for Child, Youth, and Family Policy, The Heller School for Social Policy and Management, Brandeis University; Roderick J. A. Little, Department of Biostatistics, University of Michigan; Jenae M. Neiderhiser, Human Development and Family Studies, The Pennsylvania State University; Thomas Smith, National Opinion Research Center, The University of Chicago; Deborah Klein Walker, Principal Associate, Abt Associates Inc., Cambridge, MA; Lawrence Wissow, Department of Health, Behavior, and Society, Johns Hopkins Bloomberg School of Public Health; and Robert Wright, Division of Emergency Medicine, Children's Hospital Boston.

Although the reviewers listed above provided many constructive comments and suggestions, they were not asked to endorse the content of the report nor did they see the final draft of the report before its release. The review of this report was overseen by Nancy Adler, Center for Health and Community, University of California, San Francisco. Appointed by the National Research Council, she was responsible for making certain that an independent examination of this report was carried out in accordance with institutional procedures and that all review comments were carefully considered. Responsibility for the final content of this report rests entirely with the authoring panel and the institution.

Samuel H. Preston, *Chair*
Panel to Review the National
Children's Study Research Plan

Contents

Executive Summary

On October 17, 2000, Congress enacted the Children's Health Act (Public Law 106-310). Section 1004 of the act "authorize(s) the National Institute of Child Health and Human Development to conduct a national longitudinal study of environmental influences (including physical, chemical, biological, and psychosocial) on children's health and development." In response to this act, the National Institute of Child Health and Human Development (NICHD), in cooperation with the Centers for Disease Control and Prevention, the U.S. Environmental Protection Agency, and the National Institute of Environmental Health Sciences, is planning to conduct the National Children's Study (NCS). This report is a review and assessment of the research plan for the NCS (National Institute of Child Health and Human Development, 2007).

The NCS is planned to be the largest long-term study of environmental and genetic effects on children's health ever conducted in the United States. It proposes to examine the effects of environmental influences on the health and development of approximately 100,000 children across the United States, following them from before birth until age 21. It defines environment broadly by including a wide array of measures of biological, chemical, physical, genetic, social, psychological, cultural, geographical, and other factors in a child's environment that can affect health and development. It proposes to examine the relationship between many different exposures and many different outcomes. By archiving all of the data collected, the NCS is intended to provide a valuable resource for analyses conducted many years into the future.

NICHD requested the Committee on National Statistics of the National Research Council (NRC), in collaboration with the Board on Children, Youth, and Families of the NRC and the Institute of Medicine (IOM) and the IOM Board on Population Health and Public Health Practice, to conduct a review of the research plan for the NCS. The purpose of the review is to assess the scientific rigor of the NCS and the extent to which it is being carried out with methods, measures, and collection of data and specimens to maximize the scientific yield of the study.

The panel concludes that the NCS offers an excellent opportunity to examine the effects of environmental influences on child health and development, as well as to explore the complex interactions between genes and environments. If the NCS is conducted as proposed, the database derived from the study should be valuable for investigating hypotheses described in the research plan as well as additional hypotheses that will evolve. Nevertheless, there are important weaknesses and shortcomings in the research plan that diminish the study's expected value below what it might be. This Executive Summary provides a brief overview of our assessment of the study's strengths and weaknesses; the box at the end of this summary lists the panel's recommendations for improvements to the study. Although we recognize that their implementation may raise issues of added cost and response burden, we urge that they receive serious consideration.

STRENGTHS

If the NCS is conducted as proposed, its strengths would include:

1. ***Responsiveness to the Children's Health Act of 2000*** The stated goals for the NCS, and the design of the NCS for achieving those goals, broadly reflect the stipulations of the Children's Health Act.

2. ***The large number of births to be included*** 100,000 births would provide enough statistical power to examine many hypothesized relations that cannot be investigated with smaller samples.

3. ***The longitudinal design stretching from before birth until age 21*** A data set that contains data gathered prospectively over the entire course of pregnancy, childhood, adolescence, and early adulthood will enable many new life-cycle relations between exposures and outcomes to be investigated. Data gathered prospectively (or with relatively short retrospective periods) should be more precise than data that are based on long periods of recall. A particularly attractive feature of the study is the effort to recruit births before conception and during very early periods of gestation, when certain environmental exposures may prove to be critically important.

4. ***The many variables to be generated on both outcomes and exposures*** The enormous array of social, psychological, biological, chemical, and

physical measures that will be generated under present plans will permit investigation of relationships that have not previously been studied. Some of these relationships are included among the study's hypotheses, but fortuitous and unanticipated findings can also be expected. At the same time, the large number of variables increases the risk of establishing "false positives," that is, relationships that appear to be statistically significant but have in fact been generated by chance. In addition to the variables generated by the study itself, many ancillary variables on characteristics of participants' communities from readily available sources, such as the decennial census, surveys, and administrative records, can be appended to the data set to augment the analytical usefulness of the study.

5. ***The well-designed national probability sample*** The births selected for the NCS will be identified from a probability sample of households chosen with standard and well-justified sampling techniques. The use of established random selection methods at each sampling stage will ensure that the NCS samples of households, eligible women of childbearing age, and births are national probability samples. We endorse the study's decision to use probability sampling without oversampling any groups. Statistical power could have been increased for any particular investigation by implementing a different sampling design. But modifying the sampling scheme to better address any particular relationship would often have reduced the power for investigating other relationships. The possibility of adding more precisely targeted studies—when resources and respondent burden permit—has been appropriately built into the research design. Nevertheless, it is important to note that the sample size and sampling scheme of the study represent a compromise and are not designed to address any single hypothesis.

WEAKNESSES AND SHORTCOMINGS

1. ***Absence of an adequate pilot phase*** A principal shortcoming of the NCS, as planned, is the absence of a pilot phase. The study design is extremely complex in terms of identifying subjects, enlisting their enrollment and continued participation, administering the very large number of survey and clinical instruments, and managing huge databases generated by disparate organizations. In addition, we raise many questions about the instruments that have been chosen and about the timing of their application. We think that, if the study is to achieve its promise, experimentation is needed with respect to methods to increase response rates and data instruments. Many of the concerns that we raise about the research plan could be addressed in a pilot phase.

Data-gathering will begin at the seven centers designated by NICHD as Vanguard Centers a year earlier than elsewhere. The data that will be generated in Vanguard Center sites are expected to become part of the final data set, meaning that experimentation is likely to be kept to a minimum. Moreover, the lead time between the start of data collection at the sites in the Vanguard Centers and elsewhere appears too short to take full advantage of the lessons that will be learned. The absence of a pilot phase is a serious shortcoming that could be at least partially addressed by increasing the delay between data-gathering in the Vanguard sites and elsewhere and by treating the Vanguard sites in a more experimental fashion. A delay for the enrollment phase of the study would also allow time to consider more fully the appropriate conceptual framework and specification of hypotheses and measures for the study.

2. ***Decentralization of data collection*** Data collection will be the direct responsibility of 35-40 different study centers. While the study centers will be supervised by government officials and the coordinating center, they will individually contract with outside agencies for data collection. This unusually decentralized data collection strategy reduces the chances that data will be of uniformly high quality over the life of the study and sharply increases the burden of supervision. More centralized and conventional models, such as that employed by the large National Longitudinal Study of Adolescent Health, appear more likely to produce high-quality data. Given that the decentralized approach could not readily be altered at this stage because contracts already have been signed, it will be incumbent upon the government to ensure that staff and other resources are sufficient to closely monitor data collection activities and take prompt remedial steps as necessary. A detailed and specific contractual plan is essential for this purpose.

3. ***Inadequacy of plans to maximize response rates and retention rates*** The success of the study will depend critically on the initial survey response rates and the subsequent rates of sample attrition. The NCS research plan does not explicitly address the best methods and procedures for achieving the ambitious baseline response rates that are targeted. The importance of efforts to increase initial response rates by dealing rapidly with underperforming sites cannot be overstated. Maintaining the representativeness of the sample over time is key to the quality of the results. Little is said in the research plan about how the study expects to maximize retention of sample cases. The risk of sample attrition is especially great in a study such as the NCS that targets children and young adults, the most mobile segments of the population. Typically, the largest loss to follow-up occurs in the early stages of a longitudinal study. Ascertaining the best methods to increase initial response rates and to reduce attrition rates is a matter of great urgency for project management.

4. ***Weakness of conceptual model*** The research plan does not define the basic concepts of health and development. While mentioning them, it frequently defaults to a deficit model that focuses on disease and impairment and the risk factors that contribute to them, rather than on the

factors that encourage healthy development. Late additions to the set of hypotheses that are addressed to healthy development are not well specified. Consequently, there is an imbalance of hypotheses, with specific hypotheses about disease conditions and vague hypotheses about social environments and children's intellectual and social development. The same imbalance appears in the measures selected. Little attention has been paid to outcomes in later childhood and adolescence that might have encouraged attention to additional or alternative exposures.

5. ***Weakness of certain data instruments*** The success of the study depends critically on the quality of the data instruments to be employed for assessment of environmental exposures and child outcomes. Our review revealed gaps, uncertainties, and insufficient rationales for a substantial number of instruments. Among the measures that appear most appropriate are those related to asthma and to biological exposures such as allergens and maternal physical activity. The list of proposed variables about which we have raised questions includes: birth defects, pregnancy outcomes (including fetal death), child mental health disorders, maternal depression, brain injuries, reproductive development outcomes, nonpersistent organic chemicals, pesticides, childhood infections, and the social environment in the home. The set of psychosocial measures selected appears particularly problematic. We hope that the issues we have raised about these measures will encourage a reconsideration and fresh articulation of the bases on which they were chosen, including the introduction of additional measures as substitutes or supplements when appropriate. We also suggest that efforts to assess the validity and reliability of the most problematic measures be made by the Vanguard Centers for their sites.

Even when suitable measures have been selected, the timing of their deployment—as proposed in the research plan—will leave large gaps in the measured trajectories of child health and development. These gaps, particularly with respect to in-person home or clinic visits, will make it difficult to identify critical periods of exposure to various environmental agents. The period that will be most successfully studied is that of pregnancy, birth, and the first year of life because that is the period of most intense observation. At older ages, we urge more frequent measurement of key variables, at least for a subsample of the NCS participants.

6. ***Insufficient attention to racial, ethnic, and other disparities*** The Children's Health Act asks the study to "consider health disparities among children," a phrase that typically directs attention to racial and ethnic disparities and can also include language, socioeconomic, and geographic area disparities. While the study will gather a great deal of information that is relevant to understanding such disparities, the research design was not informed by a concern with understanding their basis. The absence of virtually any hypotheses about racial and ethnic disparities is striking. In

particular, there is no attention to generating data on how individuals from different groups may interact with health systems, a factor whose importance has been suggested in many previous studies, nor on psychosocial experiences that differentiate among population groups.

7. ***Failure to adequately integrate data from medical records*** The use of health services is an important variable intervening between exposure and outcome. Given a particular exposure to environmental conditions, the quality of the medical services that are employed can make a large difference in outcomes. High-quality information about the use of medical services would also help to address major questions in health policy. These would include, as just noted, the potential sources of racial and ethnic disparities in health outcomes. Data on the use of health services will be derived almost exclusively from retrospective reports of parents, which are often unreliable. While they may be expensive to collect and mobilize and are themselves imperfect, records from physicians and hospitals would provide an extremely valuable and sometimes indispensable supplement to parents' reports. Medical records could also play an important and often central role in clarifying diagnoses and identifying patterns of child development. We urge that greater attention be paid to incorporating such data.

8. ***Failure to plan adequately for disclosure of risk to participants*** As soon as data collection begins, the NCS will face questions about the circumstances under which information about a child's health and development, as well as his or her exposure to toxic agents, should be conveyed to participants. The study plans to provide information on conditions that are "clinically relevant and actionable," but this is not a conventional concept and needs to be defined and made operational. There is insufficient detail in the research plan about how decisions will be made about what to disclose. Some of the decisions—for example, regarding transmitting information about fetal defects encountered through ultrasounds—are urgent to make. Clearer plans must also be developed regarding what parents and children need to be told about emerging *research* findings.

9. ***Failure to plan for rapid dissemination of data*** We think that the present plan is unwise in reserving a period of time for researchers associated with the data collection phase of the study to have exclusive access to its data for analytic purposes. Such a practice slows the advance of science and violates increasingly widespread norms, including those prominent within the National Institutes of Health itself. We urge wide and rapid dissemination of the data produced by the study. The data will be used for many analytic purposes that cannot presently be anticipated. The NCS is a national study paid for by public dollars, and we think that the data and results should be made accessible to the entire research community (with appropriate protections to preserve confidentiality) as soon as practicable.

Achieving rapid data dissemination will require early and elaborate organizational and budgetary planning.

CONCLUDING OBSERVATION

It is clear from our review that the NCS offers not only enormous potential, but also a large number of conceptual, methodological, and administrative challenges. In addition, funding uncertainties make it difficult to plan beyond the relatively short period for which funds have been appropriated. Like the scientists associated with the study itself, we are eager for it to succeed. We present our recommendations in the hope that, as it goes forward, the NCS will achieve its intended objective to examine the effects of environmental influences on the health and development of American children.

RECOMMENDATIONS

The panel presents recommendations in 23 areas. The recommendations are presented in the box grouped according to these areas, and keyed to the chapter in which they appear in the body of the report.

Goals

Recommendation 2-1: The NCS should give priority attention to seeking ways to bolster the ability of the study to contribute to understanding of health disparities among children in different racial, ethnic, and other population groups, including the reestablishment of a working group to oversee this area and the encouragement of appropriate adjunct studies.

Recommendation 2-2: The NCS should seek resources and develop methods to obtain more frequent in-person measures and medical and other administrative records data on study participants.

Conceptual Framework

Recommendation 2-3: The NCS should clearly define the key constructs of child health and development and more fully develop a conceptual framework for understanding child health and development over the course of infancy, childhood, and adolescence.

Using the Vanguard Centers as Pilots

Recommendation 2-4: We strongly urge the NCS to delay enrollment at new sites to make effective use of initial findings from participant enrollment and data collection in the Vanguard Center sites to improve study procedures, as appropriate, and to refine key concepts, hypotheses, and measures of outcomes and exposures. Throughout the life of the study, the NCS should use the Vanguard Centers to pilot test and experiment with data collection methods and instrumentation.

Pregnancy Outcomes

Recommendation 3-1: The NCS should consider replacing research on subclinical maternal hypothyroidism as a factor in adverse pregnancy outcomes with research on the effects of a broader set of maternal physical and mental health conditions, such as maternal depression, maternal perceived stress, and maternal periodontal disease.

Recommendation 3-2: The NCS should develop refined, detailed protocols for investigating all pregnancy outcomes, specifically a detailed protocol for obtaining information on various types of pregnancy loss, before beginning data collection at the Vanguard Centers, given that pregnancy outcomes are among the first outcomes to be examined; many outcomes lack clarity in measurement; and there are important questions regarding the adequacy of statistical power and the planned data collection (for example, the need for prepregnancy measurements of some exposures).

Neurodevelopment and Behavior and Child Health and Development

Recommendation 3-3: The NCS should develop a clearer rationale for the selection of specific neurodevelopment and behavior disorders to be considered in the study and a clearer conceptual basis for the assessment of normal child health and development trajectories and outcomes. Clarity is needed to guide the choice of outcome measures and exposure measures and the frequency and types of contacts (at the home, in clinics) with study participants in order to obtain the best information possible within resource and burden constraints.

Asthma

Recommendation 3-4: The NCS should develop a clearer rationale for its hypotheses about factors that may increase the incidence of asthma. These should focus on prenatal and early life risk factors.

Obesity and Growth

Recommendation 3-5: The NCS should reevaluate its main hypotheses to be addressed in the study of childhood obesity and consider adopting a broader approach that incorporates social and psychological factors as well as biogenetic ones. Such an approach would help the study identify the constellations of key factors and their interrelationships that are important to understand in order to develop the most effective public health measures to reduce childhood obesity.

Injury

Recommendation 3-6: The NCS should consider replacing research on repeated mild traumatic brain injury (rMTBI) with more nuanced research on other injury-related topics, such as environmental factors in childhood injuries and the effects of clinical response to injury (treatment or nontreatment).

Hormonally Active Agents and Reproductive Development

Recommendation 3-7: The NCS should develop refined and detailed protocols for studying reproductive development outcomes, which, as presented in the research plan, often lack clarity in measurement and research design. Outcomes that are measured at birth for which there is little time to refine research protocols require immediate attention. The NCS should use results from the Vanguard Centers, such as estimates of the prevalence of specific reproductive development outcomes, to assist in protocol development, and it should consider the usefulness of substudies of high-exposure population groups.

Demographic and Socioeconomic Measures

Recommendation 3-8: The NCS should add to its well-planned battery of demographic and socioeconomic measures questions on immigrant generation, languages spoken, and, if possible, the legal status of the parents and child.

Chemical Exposure Measures

Recommendation 3-9: The NCS should consider the use of personal air sampling methods for a subsample of participating women and their children for measuring exposure to air pollutants.

Recommendation 3-10: The NCS should incorporate methodology to measure paternal exposure to environmental chemicals (both persistent and nonpersistent). More generally, the NCS should consider collecting for fathers, not only chemical exposures, but also biological samples and interview data on paternal characteristics that may affect children's health and development to the same degree as it collects such information for mothers.

Physical Exposure Measures

Recommendation 3-11: The NCS should provide a clearer rationale for some of the housing and neighborhood conditions it proposes to measure and revisit its data collection plans to ensure that needed measures are obtained at developmental stages when children may be more vulnerable to risk factors. The goal should be a set of measures and data collection plans that are optimal with regard to analytic utility and response burden.

Psychosocial Exposure Measures

Recommendation 3-12: The NCS should reconsider its psychosocial measures to ensure that they will provide high-quality data for outcomes of interest for child health and development. In the face of resource and respondent burden constraints, the NCS should emphasize the quality and analytic utility of information, even if some measures must be dropped in order to substitute other assessments more desirable on various grounds.

Recommendation 3-13: The NCS should dedicate a portion of funds to support research and development of reliable and valid instruments of key psychosocial measures that are practical and economical to administer.

Biological Exposure Measures

Recommendation 3-14: The NCS should review some of the proposed measures of biological exposures, such as maternal glucose metabolism and child cortisol levels, to ensure that the proposed times for data collection are appropriate for capturing the underlying exposure.

Genetic Measures

Recommendation 3-15: The NCS should adopt a clear mechanism by which genetic association studies are internally and, optimally, externally validated before any results are published or released to the media. The NCS should also revise its proposed "established" candidate gene approach to take advantage of the new information emanating from the current wave of genome-wide association studies, with appropriate replication.

Recommendation 3-16: The NCS should consider consolidating its genetics studies in order to reduce costs and to coordinate the best science at the least cost to the project. One approach would be to simply collect the biological samples and properly store them for later genetic analysis when a better selection of polymorphisms and cost-effective genotyping across studies are possible.

Missing Exposures

Recommendation 3-17: The NCS should add measures of access to and quality of services, including medical care, education, child care, and services, as potential mediators of health and development outcomes and to improve the assessment of information obtained through maternal reports.

Data Linkage

Recommendation 3-18: To facilitate linkages of NCS data with environmental exposures from other databases, such as measures of demographics, crime, government programs, and pollution, the NCS should develop a plan for geocoding the residential addresses from prebirth through adulthood of all participating children to standard census geographic units. In addition, the study should develop arrangements by which researchers, both inside and outside the NCS study centers, can access geocodes for respondent addresses and are encouraged to perform linkages and make their environmental information available to the NCS analysis community. Such arrangements must safeguard the confidentiality of NCS respondents.

Sampling Design

Recommendation 4-1: The NCS should consider modifying the sampling design to allow for flexibility in increasing the number of study participants in the event that the estimated number of screened households needed to reach 1,000 births per primary sampling unit (PSU) is incorrect.

Recommendation 4-2: The NCS should consider the proposed household enumeration approach to be experimental and should conduct carefully designed field studies to clearly establish the statistical and practical implications of the proposed adjudicated listing approach.

Recommendation 4-3: To ensure a diverse exposure profile in the sample, the NCS should consider a careful assessment of variation in ambient exposure to chemical agents within each PSU. If the set of segments in a PSU can be classified by combined exposure to a group of important chemical agents, this information could then be used to form varying exposure-level strata for segment sampling in each PSU and thus ensure a range of ambient exposure to relevant environmental agents.

Data Collection

Recommendation 4-4: The NCS should consider ways in which the survey data collection could be consolidated into a smaller number of highly qualified survey organizations.

Recommendation 4-5: Because of the complexity of the proposed organizational model for data collection and the difficulty of maintaining the quality and uniformity of data collection procedures across a large number of study sites, the NCS program office should establish and monitor strict standards for enrollment, retention, and data collection at each of the study sites and be prepared to take immediate corrective action if sites do not meet high-quality standards in data collection.

Recommendation 4-6: The NCS should prepare a plan for monitoring progress of the study in reaching its sample size goals. As part of the plan, the NCS should take advantage of the experience of the Vanguard Centers to evaluate initial enrollment rates, the effectiveness and potential respondent burden of the interview instrument, and the ability of the Vanguard Centers to obtain the required household environmental measures reliably.

Recommendation 4-7: To resolve issues that arise during data collection, the NCS should set aside sufficient resources to maintain an ongoing program of methods research and field experimentation. Among the issues that might be addressed in this research are the reliability and validity of previously untested survey questions and measurement strategies, the effectiveness of sample retention procedures, predictors of response outcomes associated with sample initial recruitment and subsequent annual retention, error implications of unit nonresponse, adjustment strategies for unit nonresponse, and methods for dealing with item nonresponse.

Data Analysis and Dissemination

Recommendation 4-8: The NCS should begin planning for the rapid dissemination of the core study data, subject to respondent protection, to the general research community and for supporting the use of the data after dissemination. The costs of implementing this plan should be estimated and set aside in future NCS budgets. Dissemination includes not only the publication of findings through reports and scientific papers and the production of documented data files for researchers, but also active support in the use of NCS data by the broadest possible range of qualified investigators.

Criteria for Giving Information to Participants

Recommendation 5-1: The NCS should define the criteria and the process for deciding what individual clinical and research information, such as environmental assessments, test results, and survey scales, will be given to children and their families.

Protection and Release of Information

Recommendation 5-2: NCS and non-NCS investigators should be given equal access to the full NCS data as soon as they are cleaned and documented. To protect respondent confidentiality, all analyses should be performed with the kind of strict safeguards employed by the Census Bureau research data centers.

Community Engagement

Recommendation 5-3: The NCS should engage communities in selected study implementation, data analysis, and data interpretation activities that go beyond recruitment. The NCS should consider requiring every study center to formulate a more detailed plan to engage and collaborate with local communities.

1

Introduction

The National Children's Study (NCS) is planned to be the largest long-term study of environmental and genetic effects on children's health ever conducted in the United States. It proposes to examine the effects of environmental influences on the health and development of approximately 100,000 children across the United States, following them from before birth until age 21. It defines environment broadly by including a wide array of measures of biological, chemical, physical, genetic, social, cultural, geographical, and other factors in a child's environment that can affect health and development. It proposes to examine many different exposures and establish—or rule out—relationships between them with many different outcomes. By archiving all of the survey, observational, medical, and other data collected, including biological and environmental specimens, it is intended to provide a valuable resource for analyses conducted many years into the future.

BACKGROUND

Origins of the NCS

The impetus for the NCS, which has been in development since 2000, has many sources. A confluence of concerns led to the establishment, in April 1997 by Executive Order 13045, of the President's Task Force on Environmental Risks and Safety Risks to Children. The task force was charged to identify and assess the nature and extent of environmental health and safety risks to children and to recommend strategies for protecting children against environmental threats. It was cochaired by the secretary of the U.S. Department of Health and Human Services and the administrator of the U.S. Environmental Protection Agency. It included representatives of 16 departments and White House offices.

Motivating the call for the task force was evidence that certain environmental exposures early in life, such as lead poisoning and alcohol exposure in utero, compromise the health of children. At the same time, there was considerable uncertainty about the relationship of other environmental factors to what often appeared to be growing health problems among children, including asthma, autism, developmental disorders, obesity, and childhood cancers.

A task force recommendation called for legislation to mandate a large study to identify the risks to children from environmental exposures. In response, on October 17, 2000, Congress enacted the Children's Health Act (Public Law 106-310). Section 1004 of the act reads as follows (see http://nationalchildrensstudy.gov):

> **SEC. 1004. LONG-TERM CHILD DEVELOPMENT STUDY.**
>
> (a) PURPOSE.—It is the purpose of this section to authorize the National Institute of Child Health and Human Development to conduct a national longitudinal study of environmental influences (including physical, chemical, biological, and psychosocial) on children's health and development.
>
> (b) IN GENERAL.—The Director of the National Institute of Child Health and Human Development shall establish a consortium of representatives from appropriate Federal agencies (including the Centers for Disease Control and Prevention, the Environmental Protection Agency) to—
>
> (1) plan, develop, and implement a prospective cohort study, from birth to adulthood, to evaluate the effects of both chronic and intermittent exposures on child health and human development; and
>
> (2) investigate basic mechanisms of developmental disorders and environmental factors, both risk and protective, that influence health and developmental processes.
>
> (c) REQUIREMENT.—The study under subsection (b) shall—
>
> (1) incorporate behavioral, emotional, educational, and contextual consequences to enable a complete assessment of the physical, chemical, biological, and psychosocial environmental influences on children's well-being;
>
> (2) gather data on environmental influences and outcomes on diverse populations of children, which may include the consideration of prenatal exposures; and
>
> (3) consider health disparities among children, which may include the consideration of prenatal exposures.

Development of the NCS

The NCS has had a long and difficult gestation since its authorization as described below. Immediately following passage of the Children's Health Act of 2000, work began to develop the NCS in response to Section 1004 (Dr. Scheidt's presentation to the panel on September 2, 2007). An Interagency Coordinating Committee was formed that has included not only the National Institute of Child Health and Human Development (NICHD), the Centers for Disease Control and Prevention (CDC), and the Environmental Protection Agency (EPA), as called for in the act, but also the National Institute of Environmental Health Sciences.

By 2002, a federally chartered advisory group, the NCS Advisory Committee, had been established, and, by 2003, an NCS Program Office had been organized in NICHD. A large number of working groups were set up to propose hypotheses to link environmental exposure measures to child health and development outcome measures in a broad array of domains. An expert group was also established to consider an appropriate sample design for a longitudinal prospective cohort study, as called for in the Children's Health Act. These entities produced white papers and methods studies and held workshops.[1] More than 2,500 professionals in a wide range of fields contributed to the NCS development (National Institute of Child Health and Human Development, 2007:Vol. 1, Sec. 3.2).

By 2005, sufficient funding was obtained for three milestone events: (1) the development of the NCS first-stage sample, comprising 110 primary sampling units (PSUs) in 105 locations around the country by the National Center for Health Statistics of CDC; (2) a competitively bid contract to Westat to run the NCS coordinating center for data management; and (3) seven competitively bid contracts for "Vanguard Centers" to begin implementing the study in mid-2008 in select PSUs across the country. Teams from the Vanguard Centers will be the first to work in their communities to recruit participants, collect and process data, and pilot new research methods for incorporation into the full study.

Also in 2005, work began on the NCS research plan, under the general direction of the Interagency Coordinating Committee and with input from the NCS Advisory Committee, the NCS Program Office, and the Coordinating Center. Also providing input was a Steering Committee of Vanguard Center principal investigators and federal scientists that was formed following selection of the Vanguard Centers. The plan was designed for scientific peer review and public comment, by providing background on the NCS and its goals, together with descriptions of and rationales for spe-

---

[1]See http://nationalchildrensstudy.gov/research/methods_studies/; http://nationalchildrensstudy.gov/research/analytic_reports/; NICHD, 2007:Vol. 2, App. J. Not all of the early white papers and methods studies remain on the NCS website.

cific hypotheses, exposure measures, outcome measures, and sampling and data collection strategies. Work on the research plan was not completed, however, until June 2007 for the reason that, until early 2007, funding for the NCS was limited, and it was not clear that sufficient funds would be provided to let contracts in addition to the Vanguard Centers or that actual data collection could go forward.

In 2007, funding was obtained to let 22 competitively bid contracts for centers to undertake sampling, participant enrollment, and data collection in additional PSUs beginning in mid-2009 (5 of the 7 original Vanguard Centers won contracts under this procurement for PSUs in addition to their Vanguard PSUs). NICHD recently announced that contracts for another 10-15 centers will be competitively bid in 2008, as will contracts for repository and laboratory facilities.

Following review of the research plan (National Institute of Child Health and Human Development, 2007; cited hereafter in this report as NCS Research Plan) by this panel, the NICHD Institutional Review Board will complete its review of the research plan and other materials, and work will be completed on the first NCS protocol. This protocol will specify the methods and data collection instruments in detail for the first 3 years of the NCS—from enrollment (prebirth and, for about 25 percent of cases, preconception) through 24 months of age for participating children. An accompanying study manual will prescribe detailed procedures for sampling, enrollment, data collection, and data processing. These materials will be submitted for review and clearance by the U.S. Office of Management and Budget (OMB) as required by the Paperwork Reduction Act. The plan is to develop protocols and study manuals every 3 years for submission to OMB to cover data collection for participating children ages 3-5, 6-8, and so on through age 21.

THE PANEL'S REVIEW

After a long development period and uncertain funding, the NCS is scheduled to begin implementation with enrollment of participants in the seven Vanguard Centers in 2008, followed by enrollment in additional centers beginning in 2009. Prior to the completion of the detailed implementation protocols and manuals for the NCS and obtaining the requisite approvals from the NICHD Institutional Review Board and OMB, NICHD decided to request a review of the research plan.

The need for a scientific review to ensure that the study is scientifically rigorous and is being carried out with the best available methods was recognized from the early days of the president's task force. Such a review was first called for by the task force cochair, Department of Health and Human Services Secretary Donna Shalala. It is required by the NICHD Institutional

Review Board and expected by the scientific community. It is also needed to address congressional concerns. The NCS Interagency Coordinating Committee and the NICHD Program Office have consistently affirmed that there would be peer review. However, because of the funding uncertainties that delayed completion of the research plan, coupled with the pressure to get the study under way now that funding is available, the time for all of the necessary review—and, just as important, for appropriate response by NICHD—is very compressed.

The Panel Charge

NICHD requested the Committee on National Statistics of the National Research Council (NRC), in collaboration with the Board on Children, Youth, and Families of the NRC and the Institute of Medicine (IOM) and the IOM Board on Population Health and Public Health Practice, to conduct a review of the research plan for the NCS. The purpose of the review is to assess the scientific rigor of the NCS and the extent to which it is being carried out with methods, measures, and collection of data and specimens to maximize the scientific yield of the study. Topics to be addressed include proposed outcomes and hypotheses; proposed measures of environmental exposure, genetic makeup, family and community environment, and personal characteristics; proposed data collection and analysis methods; and other aspects of the study's research plan.

To address this request, the NRC appointed the Panel to Review the National Children's Study Research Plan, a group of 12 people representing a range of expertise related to the scope of the study.

The Panel's Approach

The panel's review of the NCS research plan was conducted by carrying out one pre-meeting of a subgroup of members, two in-person plenary meetings lasting two days each, numerous exchanges by e-mail and telephone, and a final one-day plenary meeting prior to sending the report through the National Academies review process.

The principal materials reviewed by the panel included the two-volume NCS research plan, received from the sponsor on June 19, 2007, and posted by NICHD on the NCS website (http://www.nationalchildrensstudy.gov) for public comment at the same time, and a supplement to the research plan received on September 17, 2007. The supplement contained some corrections, some revisions, and two additional topic areas or meta-hypotheses for the NCS (for a total of 28 meta-hypotheses, each of which comprises 2-13 specific hypotheses). The sponsor provided additional material in response to various requests by the panel. In addition, panel members and staff consulted as needed the large volume of documents on the NCS website (http://nationalchildrensstudy.gov), including the minutes of the June 26-27, 2007, meeting of the NCS Advisory Committee, which included comments by its members on the research plan.

Scope and Limitations of the Panel's Review

The panel's charge is limited to a review of the scientific rigor of the overall research plan for the NCS, as set forth in the two-volume document and supplement described above. The scope of the research plan is exceedingly broad. It presents the 28 "core" meta-hypotheses (and the specific hypotheses within each topic area) that were selected to guide the study design and data collection and analysis. It covers the wide range of exposures planned for measurement in the NCS (which include the physical environment, chemical exposures, psychosocial and biological exposures, and genetic makeup), the equally wide range of outcomes planned for measurement in the NCS (which include pregnancy outcomes, neurodevelopment and behavior, child health and development, asthma, obesity, injury, and reproductive outcomes), and the interactions among these exposures and outcomes. It also includes material on sample design and statistical analyses, ethical concerns, and management. Accordingly, all of these issues are addressed in this report.

The research plan was prepared to be readable by and informative to an interested and committed reviewer. Consequently, it is long and complex in order to cover all aspects of the NCS, yet it does not include full details of the study design and measurement approaches, full questionnaires, protocols for collection of samples and specimens and other operational aspects, or detailed implementation procedures. As noted throughout this report, in some areas the panel found that the research plan provides adequate detail for scientific review, but in other areas the panel found that the plan lacks needed detail and is a work in progress.

The organization of the research plan requires the reader to consult several sections to obtain a full picture of any single topic area. For example, material on the first listed outcome area of pregnancy outcomes is provided in the research plan in Section 1.2.1.1 (brief summary), Table 7-1, and Section 8.1 (rationale) of Volume 1, and in Appendixes A-1 and A-2/Pregnancy Outcomes of Volume 2. The reader must also consult Appendixes F-I of Volume 2 for listings of measures to be collected prepregnancy, during pregnancy, and at birth.

ORGANIZATION OF THIS REPORT

The panel's report addresses the main topics in the research plan and responds to the specific questions listed below, which the sponsor asked the panel to address—or, in some instances, the panel framed. Chapters 2-6 include the panel's assessments and our recommendations for modifications to one or more aspects of the research plan.

Chapter 2 is an overview of the NCS design and comments in broad terms on the NCS goals, conceptual framework, and selection of hypotheses. It addresses the following questions:

- Does the proposed study, as reflected in the research plan, respond adequately to the directives of the Children's Health Act of 2000, Congress, and the 1997 task force in light of the finite resources for the study and reasonable limitations on participant burden?
- Does the research plan set forth an appropriate conceptual framework for guiding the design of the study?
- Does it identify sufficient and appropriate organizing hypotheses for guiding data collection and analysis for the study?

Chapter 3 describes and comments on the concepts and measures for the children's health and development outcomes and the environmental exposures that are proposed to be identified in the study. It addresses the following questions:

- Does the proposed study design include the optimal set of "priority outcomes" in terms of their public health importance as well as scientific interest to meet the aims of the study?
- Does it include the broad range of exposures and genetic measures necessary to study the priority outcomes?
- Does it employ state-of-the-science genetic and genomic measures and analyses to enable study of interactions among environmental and biological factors that may influence the outcomes of children's health and development?
- Does it employ appropriate measurements of the outcomes, exposures, and confounders as can reasonably be done given the size, cost, and participant burden of this large, wide-ranging study?

Chapter 4 considers methodological issues for data collection and analysis. It addresses the following questions:

- Does the proposed study use appropriate sampling, estimation, and analysis methods to obtain the most useful data and results from the study?
- Does it give appropriate attention to the challenges for standardized data collection from a decentralized operation in which study centers implement sampling, enrollment, and data collection?
- Does the research plan provide information on the NCS data collection strategies to maximize initial and follow-up study response rates to ensure quality data collection?
- Does the research plan address appropriate analytic and data dissemination approaches to ensure that the study research goals are achieved?

Chapter 5 addresses ethical issues for the NCS. It addresses the following questions:

- Does the proposed study provide adequate protections for human research participants?
- Does it include plans for releasing data to the research community in a manner that maximizes what can be learned from the data, at the same time protecting the confidentiality of NCS research participants?

Chapter 6 lists the panel's conclusions and recommendations in one place for ease of reference.

2

NCS Goals, Conceptual Framework, and Core Hypotheses

The National Children's Study (NCS) was designed to respond to the requirements of the Children's Health Act of 2000. Design choices were made in accordance with the overarching goals and conceptual framework developed for the study and to reflect the data needs for analysis of 28 core or meta-hypotheses (comprising over 100 specific hypotheses) that link environmental exposures to child health and developmental outcomes.

This chapter provides an overview of the NCS design to orient the reader and convey a sense of the large scale and scope of the effort; it then presents and comments in broad terms on the NCS goals, conceptual framework, and selection of core hypotheses as outlined in the research plan (NCS Research Plan, Vol. 1, Précis, Chs. 1, 2, 4). The chapter ends with a recommendation for lengthening the time between activities of the Vanguard Centers and the corresponding activities of the other study centers. The purpose of this recommendation is to provide adequate time, not only for the NCS to learn from the experience of the Vanguard Centers with regard to study operations, but also to adequately develop key constructs and hypotheses.

As a prefatory note to our critique of various aspects of the NCS in this and subsequent chapters, we acknowledge and appreciate the enormous effort that has gone into its development. Its designers, in an environment of uncertain funding and with the need to involve many different agencies and research communities, have exhibited professionalism and strength of purpose throughout. They have often confronted difficult choices, given that no study, no matter how complex, could cover every possible exposure and outcome of interest in the same depth. Funding would never be sufficient to do that, nor would the burden on study participants be acceptable. We detail instances in which we think the particular choices made for the NCS merit reconsideration or refinement—indeed, earlier work by NCS working groups and advisers often included options that we think should be on the table but did not appear in the final research plan. Nonetheless, we remain cognizant throughout of the resource limitations for the study, as well as the design constraints imposed by some of the decisions needed to bring the NCS to its current state. At the same time, we note—as does the NCS research plan itself—the opportunity afforded by the study's length to alter course in one manner or another as appropriate, not only to take account of our comments, but also to take advantage of relevant new measurement techniques and research insights.

DESIGN IN BRIEF

The NCS is designed to be a long-running, observational panel study of a nationally representative probability sample of 100,000 births to be followed from before birth to age 21. Data will be collected on multiple exposures and multiple outcomes with repeated measures over time.

Sample Design

The sample design (see NCS Research Plan, Vol. 1, Secs. 6.1-6.4) is multistage and based primarily on households (discussed further in Chapter 4). Omitting some details, the first stage (already completed) was to draw a sample of 110 primary sampling units (PSUs) in 105 locations; the PSUs are counties or groups of counties (or parts of counties in Cook County, Illinois; Harris County, Texas; and Los Angeles County, California), each of which is expected to experience a minimum of 2,000 births during a 4-year enrollment period for the study. The second stage (implemented to date in the PSUs covered by the seven Vanguard Centers) will be to draw a sample of segments in each sampled PSU; the segments are census blocks, groups of blocks or similar subdivisions (some PSUs may be other subunits, such as school catchment areas). A third stage, implemented only in very large segments of sampled PSUs, will be to draw a clustered sample of household addresses. These stages of sampling use stratification on geographic and other variables to ensure representation of population groups (such as minorities) and areas; there is no planned oversampling or undersampling by population groups or areas.

Finally, all of the household addresses in the sampled segments (or clusters) will be visited to identify women who are eligible to participate; such women include those ages 18-44 who are not pregnant as well as all women in the first trimester of pregnancy, regardless of age. Women who are not pregnant will be recontacted several times to determine whether they have become pregnant and hence are eligible to continue in the study, with the intensity of follow-up varying depending on whether a woman is deemed to have a high or a low-to-moderate probability of becoming pregnant. It is hoped that at least 25 percent of all births in the study will be to women who were not pregnant at the time of enrollment, in which case preconception measures would be available, and that 90 percent of all births in the study will be to women who were enrolled before the end of the first trimester of pregnancy (NCS Research Plan, Vol. 1, Sec. 6.1).

The final-stage sampling, enrollment of participants, and initiation of data collection will be spread over several waves, beginning in mid-2008 for PSUs covered by the Vanguard Centers, followed by another set of PSUs in mid-2009, a third set in mid-2011, and the final set in mid-2012. During the preparation and initial sampling phases, the 35 to 40 study centers that will ultimately be part of the NCS operation will engage in outreach activities to inform local communities of the study and encourage participation (see NCS Research Plan, Vol. 1, Sec. 6.5, and Chapter 5 of this report). To identify eligible women who moved into the selected segments after the initial screening, providers of prenatal care, birthing centers, and hospitals will be visited in those segments. In each of the 110 PSUs, the enrollment period will extend over 4 years, with the goal of enrolling about 1,000 newly born children by the end of the period.

Data Collection

The data collection plan (see NCS Research Plan, Vol. 1, Sec. 6.6) is complex and uses multiple collection modes. Details of items to be collected, from whom, when, and by which method have been specified in any detail only for ages prior to birth through 24 months. A provisional schedule of contacts with study participants has been outlined for ages 3-21, but this schedule is subject to change, as are the specific items and data collection methods. The research plan argues that maximum flexibility is needed to enable the NCS to take advantage of new data collection and measurement technologies.

Table 2-1 provides summaries of the planned data collection from pre-birth through age 24 months. Locations and modes of collection include personal home visits, telephone calls to homes, and clinic and hospital visits. The mother (or primary caregiver if the biological mother leaves the child after birth) is the main respondent; the biological father or partner may also participate, although relatively little data will be obtained for that person; and, presumably, the child will be interviewed later in the study beginning at an appropriate age. Types of data collected at one or more points

TABLE 2-1 Planned NCS Data Collection from Prebirth Through 24 Months After Birth

Type and Time of Contact	Questionnaire and Diary	Biologic Samples	Clinical/Developmental Examination	Environmental Samples
		Prepregnancy		
Maternal contact only (unless otherwise noted)				
Home visit–interview	Demographics Household composition Medication use Health behaviors Housing characteristics Chemical exposures Product use Occupational exposures Diet	Blood Urine Saliva Vaginal swabs Hair	Anthropometrics Blood pressure	Indoor air House dust
Phone follow-up	Diet Chemical exposures	----	----	----

		During pregnancy		
First trimester home visit	Demographics (also P)* Household composition (also P)* Medication use* Health behaviors* Housing characteristics* Chemical exposures* Product use* Occupational exposures* Diet* Medical history (also P) Stress and social support Depression Tobacco use (P) Cognition (P)	Blood (also P) Urine (also P) Saliva Vaginal swabs Hair (also P)	Anthropometrics (also P) Blood pressure (also P) Fetal ultrasound (from medical report or clinic visit)	Indoor air House dust Drinking water Soil
Second trimester phone follow-up	Major life events Mental health update Medical update Chemical exposures update Housing update	----	----	----

continued

TABLE 2-1 Continued

Type and Time of Contact	Questionnaire and Diary	Biologic Samples	Clinical/Developmental Examination	Environmental Samples
Third trimester clinic visit	Updates on: Demographics Household composition Medication use Health behaviors Housing characteristics Chemical exposures Product use Occupational exposures Diet Medical history Stress and social support Prenatal life events Depression	Blood Urine Saliva Vaginal swabs Hair	Anthropometrics Blood pressure Fetal ultrasound	Indoor air House dust (self-collected and mailed in)
		From birth on		
Maternal (M)/child (C)/ paternal (P) contact				
Birth: At delivery, hospital	Health behaviors (M) Diet (M) Chemical exposures (M) Plans for infant feeding, sleeping, etc.	Blood (M) Urine (M) Cord blood Placenta and cord samples Heel stick (C)	Anthropometrics (C) Dysmorphology and neurologic exam (C) Digital photographs of face and anomalies (C) Chart abstraction (M, C)	----

3-month phone call	Child care Medical update (C)	Breast milk (mailed in at 4-6 weeks)	----	----
6-month home visit	Stress and social support Family process and parenting practices (also P) Health behaviors (M) Depression and cognition (M) Diet (C) Medical update (C) Medication use (C) Media exposure (C) Child care Chemical exposures Temperament (C) Tobacco use (P) Cognition(P)	Urine (C) Hair (C) Saliva (M, P) Breast milk	Anthropometrics (C) Dysmorphology exam and photos (C) Dermatologic exam (C) Social development observation (M, C)	Indoor air House dust Drinking water Soil Visual assessment of house and neighborhood
9-month phone call	Child care Medical update (C) Housing update Chemical and occupational exposures (M, C)	----	----	----

continued

TABLE 2-1 Continued

Type and Time of Contact	Questionnaire and Diary	Biologic Samples	Clinical/Developmental Examination	Environmental Samples
12-month home visit	Household composition update Family process and parenting practices (also P) Health behaviors (M) Diet (C) Medical update (C) Medication use (C) Media exposure (C) Child care Housing update Chemical and occupational exposures (M, C) Language acquisition and social interaction (C) Tobacco use Cognition (if not assessed at first trimester meeting)	Blood (C) Urine (C) Hair (C) Saliva (C) Breast milk	Anthropometrics Blood pressure Dermatologic exam Cognitive exam Motor and language assessments Social development observation (child and father, if available)	Indoor air House dust Drinking water Soil Visual assessment of house and neighborhood Noise survey
18-month phone call	Child care Medical update (C) Diet (C) Housing update Chemical and occupational exposures (M, C)	----	----	----

24-month phone call	Child care Medical update (C) Housing update Chemical and occupational exposures (M, C) Life events (M)	----	----	Indoor air House dust (self-collected and mailed in)

NOTE: P = paternal, M = maternal, C = child.
*Updates if in prepregnancy cohort.
SOURCE: NCS Research Plan, Vol. 1, Tables 6-2 through 6-4.

in time include biological specimens (blood, urine, hair, saliva, breast milk), results of physical examinations (for example, anthropometrics, blood pressure), residential environmental samples (air, dust, drinking water, soil), medical records, responses to personal and telephone interview questions, mailed-back questionnaires, diaries or logs of several types, and interviewer observations. Topics covered in interviews at one or more points in time include household composition; housing characteristics; maternal (or primary caregiver) characteristics (demographics, medication use, health behaviors, chemical exposures, product use, diet, medical history, stress and social support, depression and cognition, occupational exposures, child care); paternal (or partner) characteristics (demographics, tobacco use, medical history, cognition, mental health, parenting practice); and child characteristics (medical history, diet, medication use, media exposure, temperament, chemical and occupational exposures). There is also the potential to obtain measures of neighborhood characteristics from auxiliary data sources linked to a participant's census block group.

Following data collection at age 24 months by telephone call to the mother, a clinic visit is planned for age 3 years. Another in-home or clinic visit is planned for age 5 years and every few years subsequently—provisionally, for ages 7, 9, 12, 16, and 20 (with telephone calls presumably at more frequent intervals), although the location and timing of such visits beginning with the visit for age 5 years are not yet decided (NCS Research Plan, Vol. 1, Table 6-1).

Substudies

Substudies, outside additions to the core research, and adjunct studies (see NCS Research Plan, Vol. 1, Ch. 16) are included in the research plan to enable the NCS to serve as a platform for additional scientific research. Substudies are defined to include research that is planned in a similar manner as the planning for the core research and is supported with NCS funds but applies to only a part of the study cohort. Examples are not given but could presumably include studies of particular population groups with some added data collection. The NCS will also entertain proposals from other agencies, to be funded by the proposing body, for additions to the NCS core research plan. Both substudies and outside additions to the core research must go through the appropriate reviewing process, and approved studies will be included in the workload of all NCS study centers.

Adjunct studies are defined as studies proposed by other agencies, with outside funding, to be conducted at only one or a subset of study centers on a portion of the cohort. Adjunct studies will largely be supported with outside funding, though to a limited extent NCS funds may support some adjunct studies. Adjunct studies are implemented only with the concurrence of the involved study centers, so they are optional for the centers.

STUDY GOALS

Description

The overarching goal of the NCS, as stated in the NCS research plan (Vol. 1, Précis, pp. xv-xvi), is:

> to provide information that will ultimately lead to improvements in the health, development, and well-being of children. The primary aim of the NCS is to investigate the separate and combined effects of environmental exposures (chemical, biological, physical, and psychosocial), as well as gene-environment interactions, on pregnancy outcomes, child health and development, and precursors of adult disease. In addition to this broad purpose, the Study has several more specific goals:
>
> (1) Determine the presence or absence of effects, both harmful and helpful, related to the timing, frequency, magnitude, and duration of specific chemical, physical, biological, and psychosocial exposures in children's environments from preconception to adulthood.
>
> (2) Determine possible environmental contributions to, or causes of, specific diseases and conditions of children, including, but not limited to, prematurity and other outcomes of pregnancy, neurological and developmental disorders, psychiatric and behavioral disorders, altered physical development and sexual maturation, obesity and insulin resistance, asthma, and injuries.
>
> (3) Determine how genotypic variation and mechanisms, and the interaction of genes with environmental factors, influence disease risk and developmental trajectories in children.
>
> (4) Serve as a national resource for future studies of child health and development by providing a rich database and repository of environmental and biological samples and information that can be used to address future questions and hypotheses.

Overall Assessment

We draw one overarching conclusion about the responsiveness of the NCS goals and design, in broad terms, to the requirements of the Children's Health Act of 2000.

> **Conclusion 2-1:** The stated overall and specific goals for the NCS—and the design of the NCS to achieve those goals—largely reflect the stipu-

lations of the Children's Health Act of 2000. In the broadest terms, the NCS goals and design are responsive to the call in the act for a "national longitudinal study of environmental influences (including physical, chemical, biological, and psychosocial) on children's health and development."

The NCS represents an impressive marshalling of federal agency resources and research community input to develop and move toward implementation of the most complex and far-reaching panel study of children ever mounted in the United States. Of necessity, the NCS design is observational rather than experimental; it observes children rather than assigning them randomly to treatments, which could not ethically be done for environmental exposures—for example, children could not be randomly assigned to drink water with or without a known environmental contaminant. A consequence is that, while the NCS will often be able to establish associations between exposures and outcomes and in many cases the strength of such associations, it will not generally be able to establish causality. An exception is instances in which the NCS can take advantage of natural experiments (for example, the addition to or deletion of some highly potent substance from the food supply mandated by the U.S. Food and Drug Administration, changes in assistance program eligibility requirements adopted by some states or counties—see Chapter 3).

Associational results can be very useful to the extent that they are developed using a well-specified and theoretically grounded model that includes potential confounding factors; estimated with a high degree of precision and robustness to sensitivity tests; based on measures for all variables in the analysis that are conceptually valid and obtained with a high degree of accuracy and precision; and confirmed by the results of other studies using different methods (for example, case-control studies) or other data sources (for example, other national panel surveys using comparable measures). Subsequent chapters discuss the strengths and weaknesses of the NCS observational, prospective cohort study design in detail. Here we draw a conclusion about two major strengths of the NCS design, which are the large, nationally representative, equal probability sample design and the large numbers of variables to be collected on both outcomes and exposures over a long period of time. We draw another conclusion about four overall weaknesses of the NCS design, which are an inadequate plan for understanding health disparities, inadequate conceptualization of health and development and an overemphasis on disease and impairment, constraints on the study's ability to measure chronic and intermittent exposures, and challenges to data quality.

Conclusion 2-2: The large, nationally representative, equal probability sample design, together with the inclusion of a large number of outcome and exposure measures over a long time span, are major strengths of the NCS. In particular, the sample design is an appropriate platform for the study, considering resource constraints, the need to represent all population groups and geographic areas, and the difficulty of devising an alternative disproportionate sampling scheme that would not unduly disadvantage some groups and areas that turn out to be of analytical interest.

Conclusion 2-3: In four overarching areas, the NCS design, as represented in the research plan, is not, or may not be, optimal for achieving the goals of the Children's Health Act. These areas are:

- insufficient attention to understanding disparities in child health and development among population groups of children defined by race, ethnicity, language, socioeconomic status, and geographic area, which the act explicitly mandates;
- inadequate conceptualization of important constructs, including health and development, and an overemphasis on disease and impairment relative to health and functionality and on risk factors relative to protective health-promoting factors;
- impaired data collection schedules and types of measures to support evaluation of some of the effects of chronic and intermittent exposures on child health and development; and
- underappreciation of the challenges to obtaining the highest possible quality of data from an observational design, which include the decentralized data collection structure of the study and limitations on the frequency of home and clinic visits and on the collection of medical and other administrative records for study participants.

Overall Strengths

Sample Design

A linchpin of the NCS is the nationally representative sample design, which is important for estimating the incidence and prevalence of specific health and development conditions of concern as the study cohort matures. This design choice ensures that all population groups and geographic areas are represented in the sample, responding to the requirement in the Children's Health Act to cover "diverse populations of children." The sample design also uses equal probabilities of selection, which, by definition, means that estimates for larger population groups and geographic areas will be more precise than the corresponding estimates for smaller population groups and areas.

The drawback of equal probability sampling is the inability to oversample—and thereby improve the precision of estimates for—small population groups or geographic areas that may be of particular interest in terms of their exposures to certain kinds of environmental risk factors. Given the large number of outcomes and exposure measures of interest for the NCS, however, a design that oversampled some groups and areas might well mean that some undersampled groups and areas, which subsequently turned out to be of analytical interest, would have less precise estimates than needed for analysis. (Given a budget constraint that dictates the total number of cases, oversampling some groups must be offset by undersampling other groups, which lowers the precision of estimates for the entire sample as well as the undersampled groups.) Moreover, the very large sample size of 100,000 births planned for the NCS means that many, if not all, small population groups of interest, such as minorities, will have large numbers of sample cases. Thus, African American and Hispanic children will be adequately represented for most purposes, although smaller minority groups, such as specific groups of Asian children, may not have sufficient sample size for many analyses. On balance, we support the sample design adopted by the NCS given its resource constraints (see Chapter 4), although—as noted below and elsewhere in the report—we encourage the consideration of adjunct studies that provide additional sample for scientifically interesting groups when resources and respondent burdens permit.

Breadth of Data Collected

If carried out fully as planned, the NCS will provide a massive data-gathering platform, with the advantage of an observational frame that is both long (21 years) and wide (containing many hundreds of variables measured on 100,000 children and their caregivers). This massive frame will be a major asset in identifying relationships among exposures, mediating factors, and outcomes that would otherwise be invisible or obscured and in permitting a wide range of analyses that, over time, can take advantage of new theoretical constructs and analytic methods.

The prospective longitudinal design, in which children are followed over the entire course of pregnancy, childhood, and adolescence, should enable a large number of life-cycle relationships between exposures and outcomes to be assessed, many of which could not be investigated in depth before. Some of these relationships are posited in the study's hypotheses, but others will emerge in ways that cannot now be anticipated. A related strength of the study is the effort to recruit births prior to conception and during the first trimester of pregnancy, when some kinds of environmental exposures may prove to be critically important for some kinds of outcomes.

The very richness of the data presents analytical challenges, including the risk of asserting "false positive" relationships that appear to be statistically significant but that have in fact been generated by chance. This may be particularly problematic for studies of the association of specific genes with specific outcomes—an area in which reviews of the genetic association literature have documented large numbers of findings that could not be replicated in other studies— but also for studies of environmental exposures in which the association may be at the low level (e.g., an odds ratio less than 1.5) (see Chapter 3).

Overall Weaknesses

Health Disparities

The Children's Health Act called for the proposed longitudinal study to "consider health disparities among children," and we understand that an early NCS working group considered issues of measuring health and developmental disparities among racial and ethnic groups, although any products from this group were not available on the NCS website. A study such as the NCS has the potential power to shed light on very important issues regarding the mechanisms at work that might explain the associations between low socioeconomic status and poor health and development outcomes. The rationale for including disparities is strong given the large gaps on many health measures among population groups and geographic areas. One example is the much higher rate of low birth weight among children born to African American mothers compared with other population groups (Federal Interagency Forum on Child and Family Statistics, 2005: 32). Yet neither the NCS statement of goals nor the design, as presented in the research plan, explicitly reflects this mandate. With regard to major design elements:

- As noted, the sample is large and the sample design uses equal probability sampling because of the difficulty of specifying an alternate design that would be optimal in view of the wide range of outcomes, exposures, population groups, and geographic areas of interest. This choice means, however, that some racial, ethnic, and language minorities may have insufficient sample sizes for some analyses. Moreover, the NCS proposes to accept lower response rates in areas, such as inner cities, that are traditionally hard to survey, which will reduce effective sample sizes for disadvantaged groups relative to other groups (see Chapter 4).

- Only a handful of the more than 100 specific hypotheses in the NCS research plan directly addresses disparities, even though many outcomes of interest, such as premature birth, birth defects, obesity, and premature puberty, could benefit from explicit consideration of disparities. One hypothesis considers disparities in asthma propensities by race, ethnicity, and socioeconomic status, and another hypothesis considers disparities in schizophrenia propensities by socioeconomic status. It is not clear, however, that the NCS can adequately assess schizophrenia, given that it does not follow participants after age 21. Moreover, several hypotheses that relate socioeconomic status, ethnicity, and type of geographic area to broad developmental outcomes are so general as to lack analytical rigor and meaning (see section below on "Core Hypotheses" and Chapter 3).
- The NCS data collection plan skimps on information on the availability, use, and quality of medical, educational, and other services for study participants, which may be important mediators in producing disparate health and development outcomes (see Chapter 3). This is a consequential omission because of known differences in access to and use of services and in attitudes toward service providers among racial, ethnic, language, and socioeconomic groups. For example, disadvantaged minorities, immigrants, and poor families compared with other groups disproportionately live in areas that lack medical and other services and disproportionately distrust medical and other service providers (Institute of Medicine, 2003).

Remedying the lack of attention in the NCS to understanding the reasons for the large observed disparities in child health outcomes among population groups, beginning prenatally and extending to adulthood and beyond, should be a priority for the NCS designers. The social and structural barriers to health care, including health insurance and access to health and education services, are particularly important to address in order to understand the causes and consequences of health disparities. Many of the exposures and outcomes of concern to the NCS occur disproportionately among poor and low-income children.

There are ways to accomplish this within the current NCS design—for example, by encouraging appropriate adjunct studies that would develop hypotheses and carry out analyses of racial and ethnic disparities in child outcomes. Such studies could add sample for specific racial and ethnic groups and perhaps add data items, such as more extensive medical histories from provider records than currently planned (see further discussion in

Chapter 3). In addition, information on the spatial distribution of clinics and other service providers could be obtained from the Internet and added to the NCS to shed light on study participants' access to services, although there is no ready substitute for parental reports on use of and attitudes toward service providers.

Recommendation 2-1: The NCS should give priority attention to seeking ways to bolster the ability of the study to contribute to understanding of health disparities among children in different racial, ethnic, and other population groups, including the reestablishment of a working group to oversee this area and the encouragement of appropriate adjunct studies.

Concepts and Emphasis

The NCS research plan is much stronger on disease, impairment, and the identification of risk factors for abnormal development than it is on health, functionality, and the identification of protective and positive factors for normal development. It does not, in fact, define the key constructs of health and development. Moreover, hypotheses in the area of child health and development are very general in nature, including an additional meta-hypothesis in the September 17, 2007, supplement to the research plan on "influences on healthy development." Overall, the research plan appears deficient in its conceptualization and proposed analyses of the factors that promote healthy growth and development and that forestall or moderate the adverse effects of environmental stressors and toxins (see section below on "Conceptual Framework").

Chronic and Intermittent Exposures

In a troubling number of instances, the ability of the NCS to "evaluate the effects of both chronic and intermittent exposures on child health and development," as required in the Children's Health Act, seems impaired by the proposed schedule of data collection (see Chapter 3). Four examples that apply to prebirth and the first 24 months of age (the only period for which data collection has been fully specified to date) include:

- Glucose metabolism, which, if altered in pregnancy, is hypothesized to be a risk factor for birth defects, will be measured for pregnant women on only two or three occasions, which is insufficient to capture variations that may be important for analysis. Moreover, glucose metabolism will probably not be measured in the first few weeks of pregnancy, when much of organogenesis occurs.

- Blood samples will be collected for enrolled children only at birth and at 12 months, when more frequent collection would be desirable, not only to measure certain chronic conditions, but also to pick up episodic conditions.
- Maternal depression, known to be an important stressor for children, will be measured during pregnancy and at 6 months postpartum, but not subsequently, even though depressive episodes that are consequential for child well-being may extend beyond or occur later than the pregnancy and postpartum periods. Moreover, chronic depression is more likely to interfere with child rearing than temporary depression.
- Social development of children will be observed only at 6 and 12 months, and motor and language development will be assessed only at 12 months, which is not frequent enough to identify and categorize children by important developmental lags (or advances) or to relate lags and advances to risk and protective factors.

Budget constraints and response burden issues necessarily limit the ability of the NCS to obtain all of the detail that might be desirable at the desired frequency for all domains of interest. To ensure that sufficiently rich data are collected for important domains, we urge the NCS to do two things: first, to develop a richer conceptual framework than is presented in the research plan to guide the choice of hypotheses and specification of measures (see section below on "Conceptual Framework") and, second, to revisit each hypothesis to determine whether it is sufficiently compelling and clearly specified to retain in the study and, if so, whether adequate data are proposed for collection to permit appropriate analysis. In some instances, there will clearly be a need to add measures, frequency of collection, or both, to the schedule; in other instances, it may make sense to drop or limit the scope of a hypothesis, so that, on balance, the NCS data collection schedule is optimized subject to resource and respondent burden limits.

Data Quality

The NCS prospective cohort design requires very high quality of the data over the life of the study for its findings to be useful. It is important that the quality of the data be maintained at a uniformly high level across all study locations and all years of the study. We are concerned about the ability of the NCS to provide data of comparable quality across all study locations across time given its complex, multisite organizational structure in which some 35-40 study centers have responsibility for enrollment and data collection (see the discussion in Chapter 4).

We are also quite concerned that some aspects of the proposed data collection, which appear to be largely driven by budget constraints, will provide data of lesser quality than could and should be obtained. In addition to specific issues in particular domains (see Chapter 3) and the issue raised above about the timing of exposure measures, we generally question the relative infrequency of planned personal visits, particularly in the home, and the decision to limit the collection and coding of medical and other administrative records. Beyond the first year of life, children will not be followed up annually in homes or clinics. Mothers (or primary caregivers) will be asked retrospectively over the telephone to fill in information about important intervening events and conditions, so that many data items will be subject to recall bias and selective attrition, as well as misreporting (for example, of a medical condition or a learning problem in school).

The gaps will not be filled with service-based administrative records, which, for the most part, have been judged too expensive to acquire and abstract. Thus, medical record abstraction is planned for delivery and neonatal examination for all 100,000 study participants. Following birth, however, medical and clinical event data will be collected by a personal health record, with the parent as the primary respondent, using a diary or log format developed for the NCS. In a limited number of cases, medical records may be abstracted to confirm information regarding certain conditions. The NCS will explore the use of electronic medical records, but the research plan notes that a national, uniform system of electronic records does not now exist and may not be developed within the life of the study. No collection is planned of other types of potentially useful administrative records, such as school records or social service agency records—for example, regarding foster care or reports of domestic abuse—although procedures are being developed to collect environmental samples from day care centers and schools attended by study participants.

Service-based administrative records are not necessarily as complete or accurate as might be supposed. For example, medical records will not cover children's health problems in cases in which treatment is not sought, and the recorded diagnosis or diagnoses may not accurately reflect the underlying pathologies for insurance coverage or other reasons. Nonetheless, relying almost exclusively on retrospective parental reports on medical conditions and treatments and on such sensitive topics as domestic violence or alcoholism and failing to collect records from service providers would seem to markedly impair the quality of the data for understanding child health and development.

In sum, as a result of the decision to limit personal visits following the first year of life and to limit the collection of medical and other administrative records, data quality on health conditions and other exposure and outcome measures is likely to be lower, and it will be more difficult to identify critical or sensitive periods at which time a specific exposure might cause specific alterations of function. In this regard, the NCS is currently best designed to study relationships involving factors operative during pregnancy and the perinatal period because of the intensity and frequency of observation during this phase of child development (see Table 2-1).

Although we recognize the budgetary costs of both more frequent in-home visits and the abstraction and coding of medical and other administrative records, we urge the NCS to reconsider both decisions. Children grow and change very rapidly, and it seems imperative to have more frequent personal visits. If this is not feasible for the full study, then more frequent visits should be implemented for a sufficient subsample (ideally randomly selected) to test the effects on data quality and retention of participants and make it possible to use appropriate statistical methods to adjust the data for other participants. With regard to medical records, we urge the NCS, at a minimum, to make arrangements with providers to permit the NCS to obtain relevant records to keep in a repository to be abstracted when budgetary resources allow. We also urge the NCS to investigate the feasibility of obtaining relevant records from schools and social service agencies.

Recommendation 2-2: The NCS should seek resources and develop methods to obtain more frequent in-person measures and medical and other administrative records data on study participants.

CONCEPTUAL FRAMEWORK

Description

Chapter 2 in the NCS research plan (NCS Research Plan, Vol. 1) is labeled "Conceptual Design and Framework"; it states broad design criteria, such as assessments of multiple exposures and outcomes, repeated measures, following children through to adulthood, and generalizability to the U.S. population, on which we comment in the previous section. It also provides a very simple conceptual model, in which environmental exposures in various domains are linked to outcomes in various domains with gene expression and health care as mediating factors (NCS Research Plan, Vol. 1, Sec. 2.1, Figure 2-1).

We found Chapter 1 in the NCS research plan, which is labeled "Background," to be more informative about the conceptual thinking that underlies the study design. It provides a succinct statement of the principles underlying the study. Vol. 1, p. 1-1 states:

> The National Children's Study design rests on the principle that both health and susceptibility to disease are determined by dynamic processes that occur throughout life. Perturbations ("insults") that impact health

> states may occur any time from preconception through adult life. These insults can affect viability, differentiation of major organ systems, somatic growth, and the development of functional processes including maturation of metabolic systems. A range of determinants acting either in concert or synergistically may impact growth and development. These include the built and natural environments with their chemical and physical factors, the social environment, individual behaviors, and biological factors including genetics. Of particular importance are the earliest stages of human development, pregnancy and early childhood, when cell division, differentiation, and maturation are most rapid.
>
> These health determinants may influence development in many ways. For those with high potency when acting at critical periods of development, such as thalidomide or Accutane, severe birth defects will result in most exposed offspring. Most environmental factors, however, are not so potent. More often, factors operating at critical or sensitive periods of development will interact with other factors over the life course to raise or lower the risk of adverse health outcomes. These factors may be genetic or non-genetic. For example, accelerated weight gain during childhood is associated with increased risks of diabetes and cardiovascular outcomes later in life; this phenomenon is accentuated among children born with restricted fetal growth. . . . Only with this appreciation of the complexity of interactions among genetic and environmental factors will we be able to inform the next generation of caregivers about effective prevention and treatment to lower the burden of common chronic conditions of childhood and later-onset diseases that arise from early developmental results.

Assessment

We appreciate the acknowledgment in the NCS research plan of the complexity of child development and the complex ways in which myriad factors may interact to affect that development. However, we find the conceptual framework for the NCS to be less well developed than we would have expected for this stage of the project. While invoking a dynamic, multifactorial, life-course model, the research plan does not specify any particular model or models to guide decisions on which conditions and potential causal factors are chosen for study, the timing of data collection points, the types of data to be collected, and the overall analytical approach. There is also little attempt to capitalize on the interconnections among the conditions chosen for study.

Relatedly, we find that the research plan fails to define key constructs. While noting that the Children's Health Act mandated a study on the many environmental influences affecting the health and development of children, the plan does not provide clear definitions of "health" and "development."

There are a number of existing approaches to the conceptualization and modeling of health and development that could be used as a framework against which the data collection and analysis for the NSC could be assessed. For example, a life-course approach has been defined (Ben-Shlomo and Kuh, 2002:285) as:

> the study of long-term effects on chronic disease risk of physical and social exposures during gestation, childhood, adolescence, young adulthood, and later adult life. It includes studies of the biological, behavioral and psychosocial pathways that operate across an individual's life course, as well as across generations, to influence the development of chronic diseases.

While Ben-Shlomo and Kuh (2002) focus their discussion on chronic illness, they argue that such an approach would also apply to wider notions of health and well-being. The Institute of Medicine report (2006), *Genes, Behavior, and the Social Environment: Moving Beyond the Nature/Nurture Debate*, discusses the kinds of life-course models that are needed to understand the interactions of genetic, behavioral, and social factors as they affect human health.

A life-course model would include the identification of critical or sensitive periods when a specific exposure might cause specific alterations of function, as well as the identification of risk factors or effect modifiers later in life. These additional risk and protective factors could be further characterized as acting independently or in correlated clusters or additive "chains of risk." In addition, a life-course model could incorporate different levels of influence from the individual to the family and could include neighborhood and broader societal influences. The analytic strategy for investigating such a chain of events would be broader than a focus on logistic regression and confounders and would require multilevel modeling as well. Moreover, consideration of multiple influences on outcomes could serve to link some of the outcomes selected.

A broader conception of health and a dynamic framework for assessing health and development trajectories over time can be found in a recent report by the Board on Children, Youth, and Families (National Research Council and Institute of Medicine, 2004); see Box 2-1. This approach emphasizes optimality and functionality rather than disease and deficit and in that sense is more consistent with the World Health Organization definition of health as "positive, physical, mental and social well-being and not just absence of disease" (World Health Organization, 1946; see also Drotar, 1998; Eiser and Morse, 2001). Such an approach would lead to more of a focus than is evident in the NCS research plan on factors that protect against adverse outcomes or that mitigate the effects of risk factors; on multiple factors affecting outcomes; and on the interplay of chronic illness and development for such specific conditions as asthma (see Hayes,

BOX 2-1
A New Definition of Children's Health

The report of the Committee on Evaluation of Children's Health, *Children's Health, The Nation's Wealth* (National Research Council and Institute of Medicine, 2004), defines children's health as follows (p. 33):

> Children's health is the extent to which individual children or groups of children are able or enabled to (a) develop and realize their potential, (b) satisfy their needs, and (c) develop the capacities that allow them to interact successfully with their biological, physical, and social environments.

This definition draws upon an explanation from the World Health Organization that health is a state of complete physical, mental, and social well-being, not merely the absence of disease or infirmity; a collaborative effort of the European Union Health Monitoring Programme to develop the Child Health Indicators of Life and Development (CHILD) model; the positive health principles embraced by the Ottawa Charter for Health Promotion (1986); and the research literature cited in the report.

Key features of the new definition of children's health include three distinct but related domains (pp. 34-35):

- health conditions, "a domain that deals with disorders or illnesses";
- functioning, "which focuses on the manifestations of individual health in daily life"; and
- health potential, "which captures the development of health assets that indicate positive aspects—competence, capacity, and developmental potential."

The report describes these domains in detail and also describes ways to measure not only aspects of each domain, but also the influences on children's health, which are defined to include (p. 46):

- Children's biology
- Children's behavior
- Physical environment
    - Prenatal exposures
    - Childhood exposures
    - Home, school, and work settings
    - Child injury and the provision of safe environments
    - The built environment
- Social environment
    - Family
    - Community
    - Culture
    - Discrimination
- Services
- Policy

1997; Hobbs, Perrin, and Ireys, 1985; Stein et al., 1993; Thompson and Gustafson, 1999).

Likewise there are models of psychological development that might guide the NCS. Although the research plan cites the seminal National Research Council publication on child development, *From Neurons to Neighborhoods: The Science of Early Childhood Development* (National Research Council and Institute of Medicine, 2000), it does not reflect the conceptualization of child development and the factors affecting development that are discussed in that report. In the area of physical development, the research plan, as presented, does not incorporate the notion of physical development as a normal developmental process. Finally, the research plan does not discuss the interplay of physical growth and aspects of psychological and behavioral development (French, Story, and Perry, 1995; Molinari et al., 2002).

We are not necessarily recommending that the NSC operationalize any of these constructs of health and development completely, but such frameworks could serve as benchmarks against which the study could map what is feasible for the NSC. For example, some constructs might prove too time-consuming or expensive to measure because they require laboratory-based assessments. Others might require more frequent assessments than are possible with the resources of the overall study. In any event, the rationale for what is being done would be clear, and the gaps evident. Such gaps might very well be useful guides to what substudies or adjunct studies individual sites might conduct.

Employing explicit frameworks would not necessarily mean going back to the beginning. However, it would involve a review of the relevant theoretical and epidemiologic approaches, selection of the most appropriate with an explicit rationale for the choice, a mapping of the currently proposed conditions and measures to be collected against the framework with a description of the potential choices to be made, taking account of feasibility, costs, and response burden, and an assessment of the implications for analysis capitalizing on the power and longitudinality of the NSC. In choosing specific measures to include on the basis of the chosen framework, the NCS should consider—in addition to, or in place of, those already proposed—such indicators of child well-being and mental health as secure attachment relationships, social competence, cognitive development and achievement, optimism, and empathy. One can add to this list assessments of positive emotionality, healthy attributional style, ego-resilency, and many more. Some of this work may already have been done and be contained in background white papers, but such thinking is not evident in the research plan.

We recognize the pressures on the NCS to get the study under way, but we caution against stinting on conceptual development. We think there is time to consider more fully the appropriate conceptual framework to guide the selection of measures and analytical methods for subsequent stages of the project, given that the data collection plan is being developed in 3-year stages, as described above, and only the first stage has been fully specified thus far. Furthermore, if the NCS adopts our recommendation (see last section below) to allow more time than is currently planned between the completion of an operation by the Vanguard Centers and the implementation of that operation by other centers, then there will also be time to consider more fully the appropriate conceptual framework for the study from the earliest stage.

We understand that it is not possible—and should not be attempted—to foresee every research question of interest or to develop fully specified, complete models for every possible analysis at the outset of a long study. Indeed, an advantage of such a lengthy data collection period is that the NCS should be able to support important research that has not yet been conceived. Yet without an appropriately rich conceptual framework and well-developed key constructs, there is no assurance that the most appropriate measurements will be obtained for even those conditions that are identified for study at the outset. The NCS may want to consider the use of expert consultants to assist in the assessment and specification of appropriate frameworks, constructs, and measures to guide decisions on measures and the timing of assessments to be obtained in the study, with the Vanguard Centers used for pilot testing as recommended below.

Recommendation 2-3: The NCS should clearly define the key constructs of child health and development and more fully develop a conceptual framework for understanding child health and development over the life course of infancy, childhood, and adolescence.

CORE HYPOTHESES

Description

As indicated in the research plan (NCS Research Plan, Vol. 1, Sec. 4.2), the framing of hypotheses was considered essential to guide study planning and to ensure that important questions would be addressed. In planning the study, a standard was established that a supporting hypothesis must be developed for inclusion of measures or design elements in the NCS. To guide the selection and prioritization of outcome and exposure measures, 28 topic areas for hypotheses, referred to as core or meta-hypotheses, were developed to fulfill the study's aim of ascertaining the effect of environmental factors on the health and development of children; see Box 2-2.

The specific hypotheses in each of the topic areas identify relevant

BOX 2-2
Hypotheses Topics (Meta-Hypotheses) of the National Children's Study

- Birth defects from impaired glucose metabolism
- Increased risk of preterm birth from intrauterine exposure to mediators of inflammation
- Increased risk of fetal growth restriction, preterm birth, birth defects, and developmental disabilities in children born through assisted reproductive technologies
- Maternal subclinical hypothyroidism and neurodevelopmental disabilities/ adverse pregnancy outcomes
- Nonpersistent pesticides and poor neurobehavioral and cognitive skills
- Prenatal infection and neurodevelopmental disabilities
- Gene-environment interactions and behavior
- Prenatal and perinatal infection and schizophrenia
- Family influences on child health and development
- Impact of neighborhood and communities on child health
- Impact of media exposure on child health and development
- Social institutions and child health and development
- Influences on healthy development
- The role of prenatal maternal stress and genetics in childhood asthma
- Exposure to indoor and outdoor air pollution, aeroallergens, and asthma risk
- Dietary antioxidants and asthma risk
- Social environmental influences on asthma disparities
- Early exposure to structural components and products of microorganisms decreases the risk of asthma
- Environmental exposures interact with genes to increase the risk of asthma and wheezing in children
- Obesity and insulin resistance from impaired maternal glucose metabolism
- Obesity and insulin resistance from intrauterine growth restriction
- Breastfeeding associated with lower rates of obesity and lower risk of insulin resistance
- Fiber, whole grains, high glycemic index, and obesity and insulin resistance
- Genetics, environmental exposures, and type 1 diabetes
- Repeated mild traumatic brain injury and neurocognitive development
- Behavioral exposures, genetics, and childhood or adolescence onset aggression
- Antecedents and resiliency to traumatic life events in childhood
- Hormonally active environmental agents and reproductive development

SOURCE: National Institute of Child Health and Human Development (2007, Vol. 1, Sec. 4.2, Table 4-1; specific hypotheses in each of these areas are provided in Vol. 2, App. A-2).

environmental exposures, including physical, chemical, biological, and psychosocial factors, that affect the identified priority outcomes, including pregnancy outcomes, neurodevelopment and behavior, injury, asthma, obesity and growth, child health and development, and reproductive development. Many hypotheses also take into consideration the impact of gene-environment interactions. The criteria outlined in the research plan for the meta-hypotheses and specific hypotheses are:

- they are scientifically compelling,
- they have important public health implications,
- they are feasible to test, and
- they clearly justify the need for a prospective birth cohort study of 100,000.

Assessment

Although the panel recognizes that the framing of hypotheses is essential to guide study planning, we also think that many of the most interesting studies that are ultimately carried out with the data are likely to involve concepts and relationships that could not clearly be envisioned at the outset of the study. In our view, the NCS design, as represented in the research plan, allows for the genesis of new research ideas and hypotheses in a number of ways, including the length and breadth of the data-gathering platform, the nationally representative, equal probability sampling design, the provision for substudies and adjunct studies, and the storage of biological and environmental specimens for later analysis.

Nonetheless, the set of initial hypotheses that are developed to guide the study should be carefully selected and well framed to enable the study to collect the most appropriate outcome and exposure measures. The core hypotheses need to be well based in theory and prior research; reflect appropriate conceptual models of plausible pathways that relate exposures, mediating factors, and outcomes; and relate clearly to public health policy issues.

We are in broad agreement with the NCS's selection of the seven priority outcome areas—they represent important topics for research to inform public health policy for children. Turning to the specific hypotheses in each area, however, we think that some of them are more compelling and more fully specified than others. Some hypotheses are very broad and lack clear specification and justification, and other hypotheses are very narrow and limited in usefulness. By outcome area (see Chapter 3 for fuller discussion):

- The NCS pregnancy outcome hypotheses are well justified, with the exception of subclinical maternal hypothyroidism as a factor in adverse pregnancy outcomes, for which a compelling case is not offered either on public health or scientific grounds.
- The neurodevelopment and behavior hypotheses are not well justified in terms of why some outcomes were selected and not others—for example, it is proposed to study schizophrenia, which often does not become manifest until after the period of study (after age 21), while other behaviors that manifest themselves in adolescence, such as bipolar disorder, eating disorders, substance abuse, and others, are overlooked.
- The child health and development hypotheses are very broad: they do not articulate pathways by which such factors as the family and community environments affect or mediate developmental trajectories, and they do not clearly define the range of normal development.
- The hypotheses regarding environmental risk factors for the incidence of asthma are well developed; they represent an important and original contribution of the NCS to the field of child public health.
- The obesity hypotheses focus on maternal intrauterine environments and childhood diets as factors in obesity, to the exclusion of such domains as parental support, physical activity, food availability, schools, and TV watching.
- The injury hypotheses are problematic in several respects (for example, it will be very difficult to accurately measure repeated mild traumatic brain injury) and do not address important public health concerns related to the environmental causes of childhood injuries or the effects of clinical treatment or nontreatment.
- The reproductive development hypotheses are potentially an important and original contribution of the NCS, although the specific outcomes (for example, polycystic ovary syndrome) need to be more clearly specified.

Hypotheses are nearly missing altogether in the area of racial and ethnic disparities, as we noted above. The study will undoubtedly contribute to a better understanding of the sources of racial and ethnic, language, socioeconomic, and geographic differences in child health, but it has formed almost no hypotheses about these differences that could have informed the study design and made it more useful in this regard. Such hypotheses could have included attention to perceptions of discrimination or stigma that may affect parents' mental health and their inclination to use social services and medical facilities. Based on the research plan, this subject area appears to be largely left to the future: "The NCS will be able to address many major health disparities that currently exist in the United States, and to collect sufficient data to address others not yet recognized" (NCS Research Plan, Vol. 1, Sec. 2.2.2).

Finally, we note that no matter how well specified a hypothesis may be, considerable thought and imagination will be required to test it adequately on NCS data (see further discussion in Chapter 4). The difficulties of testing hypotheses on observational data are well recognized in the social and health sciences. Such data are rife with selection biases that can produce distorted outcomes unless great care is taken in both design and analysis. For example, what appear to be neighborhood effects should often be attributed to the characteristics (sometimes unmeasured) of individuals who live in an area. As the NCS moves from data gathering to analysis, more attention will need to be given to methodological issues in the use of observational data than is evident in the research plan.

USING THE VANGUARD CENTERS AS PILOTS

An important shortcoming of the NCS, as currently planned, is the absence of a pilot phase for thorough testing, evaluation, and refinement of study concepts, measures, instruments, and procedures. The study design is complex in every aspect, from developing appropriate constructs, hypotheses, and measures of outcomes and exposures, to identifying and enrolling participants, maintaining a high response rate over the life of the study, administering the very large number of survey, clinical, and environmental sampling instruments, and managing very large databases to which many study centers have contributed. Moreover, in this and other chapters, we raise many questions about concepts, measures, frequencies and types of data collection, and other aspects of the study.

For the study to achieve its full potential, we think that more time is needed at the outset for refinement of key constructs and measures and for full testing, evaluation, and modification, as appropriate, of study procedures. At present, it appears that little, if any, formal experimentation will be included in the work of the Vanguard Centers. Moreover, while the Vanguard Centers will begin enrollment and data collection a year in advance of other study centers, a year does not seem long enough to allow for thorough analysis of how well procedures worked in the Vanguard Centers and for improvements to be effected prior to when operations begin in the other centers.

We urge the NCS to consider lengthening the time between the start of Vanguard Center operations and the start of operations in other study centers. Although we do not think that we can make a specific recommendation about the duration of the pilot, we do think that an additional

6 months to a year between the Vanguard sites data collection and the collection of data by all the sites would be beneficial. The increase in time should be implemented not only for the first phase of enrollment through age 24 months, but also for the introduction of each new set of operations and instruments for subsequent 3-year periods throughout the life of the study. In addition, we encourage the NCS to build in experimentation with instruments and study procedures into the Vanguard operations when the extant literature is not sufficiently clear about the preferred approaches.

Recommendation 2-4: We strongly urge the NCS to delay enrollment at new sites to make effective use of initial findings from participant enrollment and data collection in the Vanguard Center sites to improve study procedures, as appropriate, and to refine key concepts, hypotheses, and measures of outcomes and exposures. Throughout the life of the study, the NCS should use the Vanguard Centers to pilot test and experiment with data collection methods and instrumentation.

3

Priority Outcome and Exposure Measures

As stated in the previous chapter, the core hypotheses of the National Children's Study (NCS) were intended to serve as guidelines for the selection of outcome and exposure measures. The elaborate NCS planning process (described in Chapter 1) led to seven priority outcome areas:

1. pregnancy outcomes,
2. neurodevelopment and behavior outcomes,
3. child health and development outcomes,
4. asthma,
5. obesity and growth,
6. injury, and
7. reproductive development outcomes.

Environmental exposure factors include the natural and built environment and the psychosocial environment. They comprise a wide range of biological, physical, chemical, genetic, social, cultural, and geographical factors. The NCS will attempt to examine many different exposures and link them in dose-response relationships with multiple outcomes. The study's geographical dispersion and the varied socioeconomic and demographic characteristics of the study population have important implications for the collection of exposure measures.

This chapter first discusses each priority outcome area, which necessarily includes some discussion of the kinds of exposures that are proposed to be associated with one or more outcomes. The chapter then reviews categories of exposure measures in more detail.

PRIORITY OUTCOMES

For each priority outcome area, the discussion summarily describes the proposed hypotheses regarding specific outcomes and associated environmental factors as presented in the NCS research plan. It then offers the panel's assessment in terms of public health significance and soundness of concepts and methodology. Each area ends with one or more recommendations.

Pregnancy Outcomes (1)

Description

The specific pregnancy outcomes identified in the NCS research plan are birth defects, prematurity, outcomes of artificial reproductive technology (ART), and outcomes of pregnancy when the woman has subclinical hypothyroidism (NCS Research Plan, Vol. 2, App. A-2, Pregnancy Outcomes). The NCS proposes to focus on altered maternal glucose metabolism and folate and vitamin supplementation as risk (or protective) factors for birth defects; the role of inflammation in the pathogenesis of prematurity; the association of ART with fetal growth restriction, prematurity, and developmental disabilities; and the relationship between maternal subclinical hypothyroidism and developmental disabilities.

Assessment: Public Health Significance

The outcomes of pregnancy clearly represent an important area for research to which the NCS could make significant contributions. If the outcomes proposed for the NCS, birth defects, prematurity, and the outcomes of ART (and subfecundity generally) are certainly of public health significance. Taken together, they account for up to 15 percent of all pregnancies. Moreover, prematurity and birth defects have proven difficult to predict and prevent (Centers for Disease Control and Prevention, 2007; Institute of Medicine, 2006). Although ART is responsible for a relatively small percentage of births (1-5 percent), it nevertheless contributes significantly to poorer birth outcomes in the United States. Thus, a strength of this section of the NCS research plan is its focus on significant public health problems.

The public health significance of maternal subclinical hypothyroidism is less clear. Limited studies suggest that unrecognized hypothyroidism during pregnancy may lead to poorer cognitive function in the child. Most cases of hypothyroidism represent autoimmune disorders, but the NCS investigators hypothesize that environmental exposures may act to disrupt the endocrine system and produce hypothyroidism, although this has not been demonstrated outside the laboratory or wildlife. The reference cited (Landrigan, Garg, and Droller, 2003) does not list the candidate exposures, and it is not clear what these would be.

Another factor potentially contributing to subclinical hypothyroidism is maternal depression with alteration of the hypothalamic-pituitary-adrenal (HPA) axis. Although maternal distress can result in alterations of hormonal function and is associated with adverse pregnancy outcomes, it is unclear that the operational pathway is through subclinical hypothyroidism. Maternal depression is certainly associated with poorer cognitive and especially behavioral outcomes in the child, but, again, the operational pathway seems more likely to be through poorer maternal-child interactions than subclinical hypothyroidism. While clinical hypothyroidism is associated with such complications of pregnancy as preeclampsia, no evidence is presented that subclinical hypothyroidism poses such a threat.

In addition, because the NCS data collection sites have been selected using equal probability sampling, the distribution of exposures to environmental agents that might result in subclinical hypothyroidism is unclear. There may not be sufficient variability, especially of high and low levels of specific agents, to permit detection of their effect. In sum, the question addressed by the posited relationship of subclinical hypothyroidism and child development outcomes represents a highly speculative chain of logic, and the importance of the problem is not entirely clear, especially since it is said to affect only 2 percent of births.

With regard to other hypotheses that could be worth evaluating in the NCS—perhaps in place of the proposed research on subclinical hypothyroidism—we make three suggestions. First, the NCS could expand the proposed study of maternal depression as a factor in adverse pregnancy outcomes: The research plan limits evaluation of its role to its effects on subclinical hypothyroidism. Second, the NCS could reconsider its decision not to obtain dental records to establish maternal periodontal disease (NCS Research Plan, Vol. 2, App. A-2, p. A2-16), or at least mount a substudy (see Chapter 2) to collect the information needed to examine the effects of maternal periodontal disease on prematurity and other adverse pregnancy outcomes.

Finally, in regard to public health significance, the current list of hypotheses does not directly address one of the most critical and enduring reproductive public health issues in the United States: the causes of racial and ethnic disparities in birth outcomes, especially the elevated rates of poor birth outcomes among African American women. The current hypotheses address a relatively narrow set of clinical concerns, although this extensive study of 100,000 births has the potential to help focus the country's intellectual attention on addressing its most fundamental issues with respect to reproductive outcomes.

Assessment: Methodological Concerns

Pregnancies without a live birth One concern about the treatment of pregnancy outcomes is that the research plan provides insufficient detail to understand how pregnancies that do not end in a live birth are to be handled. The NCS preconception and early pregnancy sampling design means that it has the potential to be one of the most important sources of scientific information on fetal loss, a critical pregnancy outcome. Many pregnancies may end in very early miscarriages, even before pregnancy testing, which could affect the ability to detect an association of early termination with inflammatory factors. In addition, prenatal diagnosis may lead to termination of pregnancies in which the fetus is assessed to be severely affected or nonviable. The research plan suggests that information on such outcomes will be sought. It will be important to employ sensitivity in data-gathering so that accurate information is obtained on these matters. The extent to which autopsy and other diagnostic materials will be obtained to ensure accurate descriptions of outcomes is unclear.

Appropriateness of the NCS design Clearly, a 20-year, longitudinal cohort study with an equal probability sample is not required to study pregnancy outcomes themselves. Most, if not all, of the pregnancy outcomes under investigation will be evident within a year or two of birth or perhaps by early school age, so that a 21-year time frame is not required to study them. Nevertheless, following up children into the later years of childhood will help track whether the impact of the reproductive outcomes persists beyond early childhood and what risk factors determine persistence and severity.

Prepregnancy exposure measures Many of the important questions about the effects of various exposures on pregnancy outcomes may require obtaining assays before pregnancy. For the hypotheses considered under birth outcomes and others in the research plan that involve birth defects, the period of concern is the first few weeks of pregnancy, when much of organogenesis occurs. In terms of the hypotheses about maternal glucose metabolism, it would seem that a more efficient design for that specific investigation would be a case-control study of women with established diabetes who are intending to become pregnant with careful attention to the periconception period alterations in glucose metabolism.

One specific hypothesis involves the role of folate and multivitamin supplementation in the prevention of birth defects. While the literature cited indicates that such supplementation may reduce the risk of a number of birth defects, it is not clear that it will affect specific defects due to impaired glucose metabolism. In addition, for many of these defects, supplementation must begin before pregnancy, which again raises the question of the size of the prepregnancy sample.

Relationships among outcomes The research plan does not explicitly consider the connections, or "crosswalks" among the outcomes selected. For example, preterm infants have twice the risk of birth defects as full-term infants, but these two outcomes are not well connected in the plan. Prematurity is considered as an outcome for ART, but ART is not considered an exposure for prematurity. It is not clear how causality will be attributed when outcomes may be associated.

Statistical power The statistical power to address some of the proposed research topics is not clear. For example, the plan indicates that 1 percent of couples are exposed to ART, which would yield about 1,000 pregnancies (1 percent). As shown in the power tables (NCS Research Plan, Vol. 1, Sec. 10.2.3, Tables 10-1 and 10-2), this is a lower figure than required for many estimates. Clearly, there will be insufficient power to examine the effect of specific types of ART on the many different types of birth defects. The situation may be even more problematic with the restriction to singleton pregnancies, as the hypotheses propose. No estimate of the number of singleton ART pregnancies is given in the research plan. The proposed data collection effort does not appear to include the number of embryos implanted. This is an unfortunate omission, because a singleton pregnancy may have different implications if it results from the implantation of a single embryo rather than from the implantation of two or more embryos with spontaneous intrauterine demise or selective reduction.

The question of statistical power also affects the hypotheses regarding altered maternal glucose metabolism and birth defects. The research plan argues the importance of this question from the rise in obesity and type 2 diabetes. Even if the prevalence doubles from the plan's estimates, only about 10 percent of women with altered glucose metabolism will have a child with a birth defect. The power calculations are based on all birth defects and all heart defects, yet "birth defects" is not a homogeneous group of conditions. Even such categories as "heart defects" comprise a large number of derangements of organogenesis. The ability to detect specific defects of a single organ, the heart, is considerably less than that suggested by the calculations for all defects. Thus, it is not clear that the study is adequately powered to explore the hypotheses under question.

Ascertaining birth defects Physical exams and digital photographs with specific attention to dysmorphology are planned for birth and 6 months to identify birth defects; however, these methods will not pick up birth defects that are not associated with external stigmata. For example, the diagnosis of heart lesions not associated with other external physical signs might be missed without such specific examinations as EKGs and echocardiograms. Presumably, significant birth defects will be reported by the mother during subsequent interviews, but this is not clear. The accuracy of maternal reports in characterizing birth defects needs to be determined. Medical records would provide more accurate diagnoses, but the NCS only plans to abstract medical records at the time of delivery and neonatal examination for both mother and infant. Subsequent to the birth, medical and clinical event data will be collected by a personal health record only with the parent as the primary respondent (see Chapter 2).

The research plan proposes ultrasound examinations in the second and third trimesters; however, the second trimester ultrasound will be obtained only if the mother has not already had an early ultrasound for gestational age dating (NCS Research Plan, Vol. 1, Sec. 6.6, Table 6-1). Such early ultrasounds may vary considerably in the quality of the reading. In addition, such limited periodicity would not be sufficient to detect many instances of fetal growth restriction.

As noted above, the teratogenicity of an exposure is often dependent on the timing of the exposure during pregnancy. Experience of a teratogen in the first few weeks of pregnancy, when major organ development is occurring, is more likely to cause greater disruption than later in pregnancy. This argument would suggest that if, for example, impaired glucose metabolism serves as a teratogen, then it would be especially important to assess exposure in the first few weeks of pregnancy, when major organs are developing. In this regard, the proposed use of hemoglobin A_{1C} as a measure will reflect the average blood glucose over weeks and not any fluctuations around the period of conception and organogenesis. Blood glucose measurements at various visits will reflect glucose metabolism at that visit and may not capture variations that occur as the woman's metabolism adjusts to pregnancy.

Interventions The degree to which the study would ascertain ART interventions other than in vitro fertilization is not clear. Superovulating agents are not only more common, but also are frequently used by practitioners who do not specialize in the treatment of infertility. However, some of the diary and hormonal information collected by ART centers will be unavailable with these techniques.

Moreover, ART is but one of many interventional reproductive health service technologies that could be a focus of this large perinatal study. For example, there is critical debate and a serious lack of information about the sequelae of Caesarian births and the use of analgesics. Similarly, significant variations in outcomes among hospitals with neonatal intensive care units have been documented (Vohr et al., 2004). Attribution of developmental outcomes to ART or to any other prenatal intervention or exposure needs to account for this variation.

Recommendations

We have identified a number of issues and concerns with the NCS research plan proposals for assessing environmental influences on pregnancy outcomes. We offer two recommendations for steps that we judge to be of high priority for the NCS: one on the set of hypotheses that merit study and the other on the need for more specificity of the proposed research on pregnancy outcomes, which will be among the very first for which data are to be collected.

> **Recommendation 3-1: The NCS should consider replacing research on subclinical maternal hypothyroidism as a factor in adverse pregnancy outcomes with research on the effects of a broader set of maternal physical and mental health conditions, such as maternal depression, maternal perceived stress, and maternal periodontal disease.**

> **Recommendation 3-2: The NCS should develop refined, detailed protocols for investigating all pregnancy outcomes, specifically a detailed protocol for obtaining information on various types of pregnancy loss, before beginning data collection at the Vanguard Centers, given that pregnancy outcomes are among the first outcomes to be examined; many outcomes lack clarity in measurement; and there are important questions regarding the adequacy of statistical power and the planned data collection (for example, the need for prepregnancy measurements of some exposures).**

Although development of a detailed protocol will not help the statistical power issues for some outcomes as now specified, detailed protocols with more specific calculations on anticipated numbers of various exposures and subjects would certainly clarify what outcomes could be realistically studied, and whether changes to the research plan might be needed (e.g., not restricting assessment of the effects of ART to singleton births). Such protocols might also indicate subgroups for which more intensive study might be warranted, such as mothers with preexisting problems with glucose control. In addition, such protocols might lead to more focused outcomes (e.g., specific definitions with prevalence) and identify additional data collection required to ascertain these outcomes (e.g., echocardiogram data on congenital heart defects).

Neurodevelopment and Behavior (2) and Child Health and Development (3)

The Children's Health Act mandate to "investigate basic mechanisms of developmental disorders" and "incorporate behavioral, emotional, [and] educational . . . consequences" of environmental influences encompasses an exceedingly broad range of developmental outcomes. Moreover, while developmental disorders are a clear study priority, environmental influences can also affect the much broader spectrum of age-normative developmental functioning. Given resource and burden limitations, the NCS faces difficult choices regarding the type and nature of its measurements of disorders and normative developmental outcomes.

To organize discussion of the issues and because the two domains are intertwined conceptually and in NCS planning, this section first discusses the specific plans for each domain and then provides a combined assessment.

Description: Neurodevelopment and Behavior Outcomes

The NCS proposes to focus on identifying specific developmental, behavioral, or mental health disorders, including sensory, motor, and learning disabilities, autism spectrum disorders, attention deficit–hyperactivity disorder (ADHD), anxiety disorders, depression, and schizophrenia and relating them to specific environmental exposures. The NCS will examine four broad hypothesized relationships (meta-hypotheses, NCS Research Plan, Vol. 2, App. A-1, p. A1-2; see App. A-2, Neuro/Behavior, for specific hypotheses within each meta-hypothesis):

> Repeated, low-level exposure to nonpersistent pesticides . . . in utero or postnatally increases risk of poor performance on neurobehavioral and cognitive examinations during infancy and later in childhood. . . .
>
> Prenatal infection and mediators are risk factors for neurodevelopmental disabilities such as cerebral palsy and autism. . . .
>
> Exposures to adverse psychological, chemical, and physical environments and other stressors during vulnerable periods of pregnancy and early childhood can interact with genotype to cause or modulate behavioral problems in childhood. . . .

> Prenatal infection and mediators of inflammation during pregnancy and the perinatal period are associated with increased risk of schizophrenia.

According to the research plan (Vol. 1, Ch. 8, pp. 8.5-8.11), the NCS will rely on a combination of screening instruments and diagnostic information to identify developmental and mental health disorders. However, as stated earlier, the National Institute of Child Health and Human Development (NICHD) study staff have indicated that resources are not available at this time to abstract medical records except at birth (see Chapter 2).

Sensory, motor, and learning disabilities Some sensory and motor difficulties are evident very early in the child's life. Learning disabilities, however, are often not identified until children enter school. Routine infant hearing screening is recorded in the hospital chart at birth, which will be abstracted by the NCS. Screening for sensory and motor disabilities will begin before the neonate has been discharged from the hospital by using the Network Neurobehavioral Scale to assess the infant's neurological status.

The NCS plans to track children's developmental status during infancy with regard to cognitive, motor, and language delays using multiple assessment strategies. At 12 months, the NCS will administer three of the Bayley III Scales of Development: Cognitive, Motor, and Language to all enrolled children to assess the achievement of developmental milestones in these domains. Actual diagnosis of learning, sensory, and motor disabilities will be confirmed whenever possible through the child's medical records, including the diagnoses and treatment plans of their medical providers. The child's health care visits will be reviewed at every contact with the parents, including both in-person contacts at 6 and 12 months and phone contacts at 3, 9, 18, and 24 months, and they will continue to be assessed regularly after that. Throughout childhood and adolescence, the child's developmental status with regard to cognitive, language, and motor functioning will continue to be assessed periodically through direct testing by the NCS and diagnoses confirmed whenever possible through health care providers.

Autism spectrum disorders These disorders are not generally diagnosed until the child's second year or later. The NCS will begin screening for autism spectrum disorders when the child is 18 months old and continue to screen for symptoms periodically through the toddler and preschool period by using the Modified Checklist for Autism in Toddlers (M-CHAT), a parental report instrument presumably provided over the telephone. The M-CHAT, however, is a screen for risk of autism; it does not yield a diagnosis of autism spectrum disorders. For diagnostic information, the NCS proposes to use diagnostic assessments conducted by the children's health care providers and abstracted from medical records whenever possible.

Behavioral, attention, and mood disorders These disorders are rarely diagnosed in infants. At 12 months the parent will be asked to complete the Brief Infant-Toddler Social and Emotional Assessment (BITSEA), a screening instrument that assesses risk for mood problems, behavior problems, and self-regulatory deficits. The BITSEA, or an age-appropriate modification of the BITSEA, will be repeated through the toddler and preschool period to track any problems over time. As the children become older, other similar screening instruments will be used, such as the widely used Strengths and Difficulties Questionnaire, which assesses conduct problems, emotional problems, hyperactivity and inattention problems, and relationship problems, and can be completed by parents, teachers, and in the teen years by the adolescents themselves.

Early diagnoses of disorders will be confirmed whenever possible through the children's health care providers' records. Later in childhood, measures and diagnostic interviews, such as the Preschool Age Psychiatric Assessments (PAPA) interview or the National Institute of Mental Health Diagnostic Interview Schedule for Children (NIMH-DISC-IV), may be used to supplement diagnostic information from children's health care providers and ensure diagnostic information on children who do not visit health care providers regularly.

Schizophrenia This psychotic disorder, believed to have both genetic and environmental etiology, will also be studied in the NCS. Schizophrenia, however, is rarely diagnosed before late adolescence or early adulthood. The research plan identifies no specific screening or diagnostic tools, stating that screening and diagnostic procedures for schizophrenia are likely to continue to evolve before the NCS children reach the life stage when schizophrenia is usually diagnosed.

Description: Child Health and Development Outcomes

Normative child health and development is concerned not with disorder or symptoms of disorder, but with individual differences in trajectories of normal, healthy adaptation over time. The NCS proposes to examine cognitive and language development and also social and emotional development. The research plan spells out five broad meta-hypotheses (NCS Research Plan, Vol. 2, App. A-1, p. A1-3; see App. A-2, Health/Development, for specific hypotheses within each meta-hypothesis).

The first meta-hypothesis links family resources and processes to the structure and quality of children's home, child care, school experiences, and economic opportunities, which, in turn, affect developmental and health trajectories. The second hypothesis links geographic area of residence to exposure to social, physical, psychological, and environmental factors that adversely affect the risk of health problems and decrease access to protective resources. The third hypothesis links media use and content (TV, video, electronic games, Internet, mobile devices) to developmental trajectories from prosocial to antisocial behavior. The fourth hypothesis links interactions between children and families and the formal child care, school, and religious institutions in their communities to cognitive, social, and emotional development.

Cognitive and language development will be tracked throughout childhood using the procedures and instruments outlined for sensory, motor, and language disabilities above. The intent in this instance will be to identify normal development.

Under the rubric of social and emotional development, the NCS proposes to cover several domains of child functioning, both intrapersonal and interpersonal, including temperament, mother-child interaction, and relationship skills. The research plan posits that assessing temperament early in development is important, as temperamental qualities not only exert direct influence on children's adjustment, but also influence parental reactions to the infant's signals and needs and thus affect subsequent development indirectly.

When the infant is 6 months old, the NCS proposes to collect maternal reports of child temperament using three subscales of the Rothbart Infant Behavior Questionnaire-Revised (IBQ-R), including activity level, fearfulness, and positive anticipation of and approach to novelty. Also at 6 months, the NCS will conduct its first videotaped observation of mother-child interaction. At 12 months, the child's social and emotional development will be assessed using parental reports on the BITSEA. During the toddler and preschool years, the same constructs will be assessed again, using the same procedures when appropriate, or using assessments that are age-appropriate measures of these constructs, such as the Strengths and Difficulties Questionnaire, which assesses prosocial behavior and relationship skills. As the child ages and begins to spend time in the broader social contexts of school and community, assessments of developmental trajectories in social and emotional competence will be tailored to include these experiential changes.

Assessment: Public Health Significance and Conceptual Concerns

Both the domain of neurodevelopment and behavior disorders and that of normal child health and development are clearly important areas to which the NCS could make significant contributions. However, the research plan has serious conceptual limitations in each domain that warrant concern.

Neurodevelopment and behavior In this domain, it is not clear by what process and reasoning certain specific outcomes were selected for study and not others. In this domain, the NCS proposes to investigate sensory, motor, and learning disabilities, autism spectrum, ADHD, depression, anxiety, and schizophrenia. One could also add such outcomes as eating disorders, bipolar disorder, suicide, substance abuse, binge drinking, and high-risk sexual behaviors—all high-risk behaviors of public health significance that are common in adolescents and many of which are likely to be influenced by developmental and environmental conditions prevailing much earlier in childhood. It is not the case that these disorders are excluded because they do not become manifest until adolescence, since schizophrenia is included and it is not typically diagnosed before young adulthood. The research plan does not indicate why the specified outcomes were chosen and other important outcomes were not. Clearly, not all disorders can be studied in the same depth, and it is important to have a clear rationale for selecting the specific ones to study.

The hypotheses that propose to link environmental exposures to adverse neurodevelopment and behavioral outcomes reflect a biomedical approach that appears limited in its ability to contribute to understanding of this complex area. For example, the hypothesis to link repeated, low-level exposure to nonpersistent pesticides to poor neurobehavioral and cognitive skills is important because of widespread levels of exposure and the likelihood that children are more sensitive to exposures than adults. However, all other factors that could mediate a relationship of pesticide exposures to adverse outcomes are lumped into "potential confounders" without a clear discussion of likely pathways of interactions that should be explored, or of likely dose-response relationships. The only exception comprises two specific hypotheses in this area that propose to examine the mediating effect of genetically decreased paraoxonase activity (an enzyme that protects against the adverse effects of low-dose organophosphate pesticide exposures).

Similarly, the proposal to study depression and other mood disorders reflects an important area of investigation. The NCS takes a largely biomedical approach to these outcomes. Specific hypotheses focus on the role of genetically based levels of neurotransmitters, such as serotonin, norepinephrine, and dopamine, as they interact with "life stress" to the exclusion of other influences and pathways that may be operative. For example, negative attributional style is widely hypothesized to be relevant to the development of depression, but there are no plans to assess this construct (Gibb et al., 2006). (Exposure to prenatal tobacco smoke and "maternal insensitivity" are posited as mediating factors in the relationship of dopamine to ADHD.)

Positive health and development outcomes In this domain, the discussion in the research plan does not define clearly how the NCS understands the key constructs of what constitutes "health" and "normal development" nor the relationship of normal variation to diagnosable problems. Moreover, the specific hypotheses in this area not only are very general regarding developmental processes and the factors that are assumed to promote healthy development, but also are not well integrated into an overall concept of healthy development. For example, under the first meta-hypothesis, the NCS proposes to look separately, and also interactively, at the effects of the nature and stability of family structures, families' social networks, family socioeconomic status, family members' health status, parents' promotion of healthy lifestyle behaviors, and parental monitoring of children's activities at home and in their neighborhoods. Similarly, under the fifth meta-hypothesis (added in September 2007), the NCS proposes to look separately at the effects of parenting behavior, child care experiences, access to and use of health and social services, and such child characteristics as intelligence and temperament, on healthy child development. Pathways for the separate and interactive effects of these factors are not clearly specified, nor are the factors always clearly defined.

Assessment: Methodological Concerns

In the neurodevelopment and behavior domain, once disorders are selected for study, they must be precisely coupled with procedures to screen for them and subsequently to diagnose them in a rigorous fashion. The current plans for achieving this goal do not appear to meet this standard. As noted above, the research plan identifies several recognized tests for screening for the disorders studied, but these screening instruments are not useful as diagnostic instruments. The NCS appears to rely mostly on health provider records whenever possible and on mothers' reports for diagnoses. Yet, as we have noted previously, plans for the use of pediatric medical records, except at birth itself, appear to be on hold because of resource constraints. Discussion of medical records with parents, which the research plan mentions, is not likely to yield reliable information. Nor does the plan address the adequacy of medical records themselves, which may be insufficient for diagnostic purposes even if they are obtained. Reliable diagnosis using *Diagnostic and Statistical Manual* criteria, subsequent to positive findings from screeners or medical records, may take extensive interview time and necessitate the gathering of denser observational data than is currently contemplated. Moreover, access to medical records information will depend heavily on the participant's type of insurance for mental health services. Developmental disabilities fall under mental health codes, so that funding may be limited and professionals willing to take such funding not readily accessible.

Moving to assessment of normal, healthy functioning, there is little attention to evaluating normal variation in such aspects as attention, working memory, executive function, language ability, attachment, and peer relations. The outcome measures discussed partially overlap with those listed in the neurodevelopment and behavior section and include:

- 3 Bayley scales;
- CDI (Communicative Development Inventory) at 12 months;
- intelligence tests—perhaps Woodcock-J scales at later ages or Bayley or Kaufman scales;
- unspecified assessments of executive function and attention;
- Rothbart IBQ-R measures of temperament;
- videotaped mother-child interaction at 6 months; and
- BITSEA.

This list appears ad hoc and not well informed by contemporary child development research or well focused on specific hypotheses or domains of interest. For example, Achenbach's measures of psychosocial functioning seem to be missing, even though they are the usual ways of looking at internalizing and externalizing behavior across a normal range. As another example, there is no plan to use the Strange Situation at 12 months to evaluate attachment security, even though there is a vast literature on attachment theory and on the prospective implications of Strange Situation classifications. Other examples of missing measures include the Continuous Performance Test (also relevant to ADHD diagnosis) and measures of verbal and visual-spatial working memory. Better measures of infant cognition would include the Fagan test, Rose's cross-modal matching paradigm, or Haith's visual evoked potential (VEP). These tests are known to predict later IQ better than the Bayley and to be sensitive to the effects of environmental contaminants when the Bayley is not (Jacobson and Jacobson, 1996). A final example is that the CDI will not, at 12 months, provide a good assessment of language acquisition. It needs to be repeated at subsequent ages and coupled with a more intensive battery at about 4 years, using, for example, the Reynell or the deVilliers test (Seymour, Roeper, and deVilliers, 2003a, 2003b).

For both outcome sections, there is no information on the reliability of tests and procedures, the length to administer them, their validity and especially their discriminant validity, when and how often they will be given, and whether there is a need to worry about repeated testing effects. These are basic research issues that should have been discussed. Finally, the research plan needs to address ethical issues regarding referral for treatment when disorders are discovered and what to do when such treatment is not covered by insurance (see Chapter 5).

In summary, in measuring neurodevelopment and behavior and child health and development outcomes, there is nearly exclusive reliance on screeners, brief questionnaires, and psychometric tests, chosen in a way that does not seem closely related to hypotheses regarding domains of interest. Direct behavioral assessments of children are very limited; instead, measures are largely being derived from reports by parents rather than from validated behavioral tests that are available. As noted in Chapter 2, there are infrequent home or clinic visits after age 2, which seriously limits the ability to observe and measure normal, or pathological, behavioral development.

We recognize that a number of well-validated instruments that are not proposed in the research plan may be too burdensome for respondents and too costly to administer given the wide range of exposures and outcomes that the NCS is striving to measure. As multifaceted as the NCS is, it cannot cover all domains in the same degree of detail, and it may be that the domains of neurodevelopment and behavior and child health and development are ones that must be left less well covered than other domains. To the extent that is the case, it is incumbent upon the NCS to carefully review the specific outcome measures that are proposed to be sure they are the best possible measures within the constraints of resources and respondent burden. In this regard, the NCS could make more extensive use of measures used in the National Childcare Study (updated as appropriate), as a large group of inventive and dedicated behavioral scientists faced the same basic constraints as the NCS when it developed the set of cognitive and behavioral measures it employed.

Recommendation

Our discussion has identified a large number of concerns. Most important is the need for the NCS to justify the selected outcomes of interest and related exposure measures, given the wide range of possible outcomes the NCS could study in the two domains of neurodevelopment and behavior and normal health and functioning. The NCS needs to indicate clearly why some outcomes were chosen and others not, by using the criteria of the research plan itself, which include the current state of knowledge, the need for a large sample size to answer key questions, and the public burden of particular disorders. Despite the wide mandate of the Children's Health Act, it is not possible to assess all normal variation in healthy functioning as well as all possible disorders in a rigorous fashion. One strategy might be to select a smaller number of disorders and orient screeners, normal assessments, and diagnostic interviews around those areas of functioning. That may allow for more principled decisions about ages at assessment, laboratory versus home visits, and other methodological issues.

Recommendation 3-3: The NCS should develop a clearer rationale for the selection of specific neurodevelopment and behavior disorders to be considered in the study and a clearer conceptual basis for the assessment of normal child health and development trajectories and outcomes. Clarity is needed to guide the choice of outcome measures and exposure measures and the frequency and types of contacts (at the home, in clinics) with study participants in order to obtain the best information possible within resource and burden constraints.

Asthma (4)

Description

The NCS proposes to examine six hypotheses about risk factors for childhood asthma (NCS Research Plan, Vol. 2, App. A-2, Asthma). Broadly, the hypotheses assume (1) prenatal maternal stress as a risk factor, mediated by genetic and environmental factors that influence immune development and lung growth/airway inflammation in early life (for example, prenatal exposure to tobacco and air pollution); (2) indoor and outdoor air pollution and aeroallergens as risk factors, mediated by genotype; (3) antioxidants in diet as protective factors; (4) social environmental factors that influence exposure to physical environmental risk factors, psychosocial stress, and adverse health-related behaviors, thereby explaining, at least in part, disparities in asthma prevalence and severity by race and socioeconomic status; (5) early exposure to common indoor aeroallergens and microorganisms as protective factors; and (6) significant gene-environment, gene-gene, and genotype-phenotype interactions as risk factors.

Assessment: Public Health Significance

Asthma is of high public health importance in terms of both the burden of this chronic disease and its documented increase in incidence. Asthma continues to be the most common chronic disorder among children in the United States, affecting 9.9 million children (or 14 percent of all children) under age 18 (Bloom and Cohen, 2007). Asthma is one of the leading causes of school absenteeism, accounting for 12.8 million missed school days in 2003 (Akinbami, 2006). It is appropriate, and responsive to the Children's Health Act of 2000, for the etiology of asthma to be a major focus and aim of the NCS.

The major strength of the proposed NCS in the area of asthma is the prospective cohort design, which will allow for in-depth investigation of early life risk factors for the development of asthma. There are only a few large prospective cohort studies on potential risk factors that contribute to the increased *incidence* of asthma. Alternatively, epidemiological studies on risk factors for the *exacerbation* of asthma are numerous. Therefore, the major strength of the NCS is the ability to study asthma incidence beginning with prenatal exposures. Follow-up into the later years of childhood is necessary to understand whether asthma symptoms persist beyond early childhood and what risk factors determine persistence and severity.

The proposed measures of asthma outcome are valid and feasible to collect. They include questionnaire data, physician-confirmed diagnoses, and such objective tests as spirometry and peak flow measures.

Assessment: Methodological Concerns

Issues of statistical power arise for research on asthma as for other outcomes. For instance, it is not clear whether there will be, for analysis purposes, (1) a sufficient number of children from a variety of housing conditions, such as inner city housing developments compared with more rural housing or (2) a sufficient number of children from a variety of geographic conditions that contribute to differences in environmental exposures, diet, and other factors. Insufficient numbers of children on these dimensions could weaken the ability of the NCS to estimate robust associations that suggest causal relationships (see Chapter 4 for further discussion).

Throughout the asthma section, the research plan refers to the collection of biological samples to measure biomarkers of exposure and response. It is not clear, however, whether the study has adequately considered the appropriate timing (during pregnancy and childhood) of the collection of these samples. An important aim of the study should be to determine whether the difference in timing of an exposure affects the risk of the development of asthma. Given the infrequency of follow-up after birth proposed for the NCS (see Chapter 2), it is unlikely that these questions of timing will be satisfactorily addressed.

The research plan proposes to measure exposure to indoor and outdoor aeroallergens, which is appropriate given the results of a recent study by researchers at the National Institute for Environmental Health Sciences, which found that more than one-half of the current asthma cases in the country can be attributed to allergens (Arbes et al., 2007). The research plan needs to provide more detail, however, on how aeroallergen exposure will be assessed and which aeroallergens the study will focus on. Aeroallergen exposure assessment is not simple and straightforward, and detailed protocols are necessary. Dust samples will be collected to measure mold, endotoxin, and common environmental antigens. It is not clear whether or how environmental chemicals, such as pesticides, and metals will be measured in the dust samples.

Accompanying the increase in asthma incidence among children has been a parallel increase in children's allergies, including severe food allergies. It would be straightforward to collect information on food allergies among the NCS cohort and to explore potential risk factors ranging from maternal diet to early life exposure to allergens. Furthermore, data suggest that children with certain food allergies, such as to eggs, have a higher risk of developing asthma. This hypothesis could readily be explored.

Recommendation 3-4: The NCS should develop a clearer rationale for its hypotheses about factors that may increase the incidence of asthma. These should focus on prenatal and early life risk factors.

Obesity and Growth (5)

Description

The NCS proposes to investigate several ways in which environmental factors may increase the risk of obesity and insulin resistance (type 2 diabetes) in children (NCS Research Plan, Vol. 2, App. A-2, Obesity/Growth). The NCS has chosen four biomedically oriented meta hypotheses to investigate: (1) impaired maternal glucose metabolism during pregnancy is directly related to the risk of childhood obesity and insulin resistance; (2) intrauterine growth restriction is associated with risk of central-body obesity; (3) breast milk compared with infant formula feeding is associated with lower rates of obesity; and (4) consumption of a high glycemic load diet is associated with obesity. In addition, the study proposes to relate the development of type 1 diabetes to an interaction between genetic susceptibility, early exposure to viral infections, and early exposure to cow's milk protein or other dietary components. The NCS research plan does include getting growth measures as direct observations. In addition, the NCS will be obtaining related blood samples for lipids and glucose metabolism abnormalities.

The NCS research plan, however, does not define precisely or provide rationale for what it means by obesity. Recent clinical recommendations focus on specific cutoff values for body mass index (BMI). However, there is controversy as to what standards should be used for BMI (Cole et al., 2000) and whether all standards apply to specific groups of children. In addition, other measures of excess fat (e.g., waist circumference, bioelectrical impedence) are available (although again, there is controversy as to the information added by such measures) (Krebs et al., 2007). The NCS should be specific as to why certain measures are preferred. There is also reference in the appendix to "body habitus," but this is likewise not defined. Exercise is not mentioned in the hypotheses on obesity and growth. It could be asked directly of parents, and it could also be examined in terms of the availability of recreational facilities in geographical areas. As with other outcomes, no mention is made of diagnostic assessments to eliminate causes of obesity that are not of interest to the NCS.

Assessment: Public Health Significance

Childhood obesity has been increasing steadily, particularly during the past two decades (Ogden et al., 2002, 2006). This rising rate of obesity is a major health concern, both because of its impact on childhood health and because of the strong association between obesity and cardiovascular disease risk, hypertension, and type 2 diabetes that begins in childhood and continues into adulthood (Institute of Medicine, 2007). Numerous studies over the last decade have indicated a large variety of interrelated risk factors for childhood obesity (Barlow, 2007; Hawkins and Law, 2006; Skelton, DeMattia, Miller, and Olivier, 2006). It certainly deserves to be a priority outcome of the NCS.

Assessment: Conceptual and Methodological Concerns

The large sample size of the NCS provides an important opportunity to evaluate such hypotheses as the effects of impaired maternal glucose metabolism and intrauterine growth restrictions on childhood obesity. In general, however, the hypotheses selected by the NCS appear too narrowly focused to result in new information that could ultimately reduce the epidemic in childhood obesity and curtail its adult sequelae. There are too many additional factors that play an important, if not essential, role in the development of obesity, including a range of social and psychological factors acting at both the individual and collective levels.

The current hypotheses selected focus on only a very small part of the complex pathways to the development and chronic condition of obesity. A more integrated and systematic examination of a wider range of factors, many of which will be collected as a part of the NCS, would be likely to have more concrete benefits for improving children's health. Such an examination should elucidate the web of physical, behavioral, social, and genetic factors that act additively or interactively to determine intrauterine growth retardation, breastfeeding versus infant formula feeding, and children's diet, along with the role of these factors during childhood development. A systems approach to understanding this complex chronic condition could be one alternative.

Although it is valid to test hypotheses about particular mechanisms of action in epidemiological cohorts, one must remember that association is not causation. For the hypothesis about glycemic load, the difficulty of inferring causation is magnified by measurement issues. There are serious concerns about the ability of the NCS to measure the consumption of high glycemic load diet using food diaries. Moreover, the time frame for sampling information about the child's diet is neither well motivated nor connected to the hypotheses to be addressed.

The research plan raises other questions. First, it clearly delineates a work-up to rule out other reasons for obesity. However, the research plan does not indicate what maneuvers will be done to ensure that the obesity detected is due to diet or physical activity and not to metabolic problems. Second, there are questions about what will be done when the obesity is detected. Will treatment plans be put into effect? How will that alter the "natural history" portion of the study?

One last major concern is the tacit, but evident, focus on the mother as the source or cause of obesity in her children. While there is little debate that maternal factors play a role, many other factors and circumstances can be expected to have an impact as well.

Recommendation

Recommendation 3-5: The NCS should reevaluate its main hypotheses to be addressed in the study of childhood obesity and consider adopting a broader approach that incorporates social and psychological factors as well as biogenetic ones. Such an approach would help the study identify the constellations of key factors and their interrelationships that are important to understand in order to develop the most effective public health measures to reduce childhood obesity.

This is another situation for which consultants could be useful to assist the Vanguard Centers in developing and testing protocols that could lead to improved instruments.

Injury (6)

Description

The NCS singles out repeated mild traumatic brain injury (rMTBI) for attention in the broad field of unintentional childhood injury, positing that it has a cumulative adverse effect on neurocognitive development (NCS Research Plan, Vol. 2, App. A-2, Injury). Under the category of injury, the NCS also proposes to examine the relationship of biological, physical, and psychosocial components of the environment and environment-gene interactions on early onset and continuation of antisocial physical aggression, as well as the relationship of genetic, family structure, and neighborhood and community factors interacting with traumatic life events on the risk of anxiety disorders.

Assessment: Public Health Significance

Childhood injuries are a major source of child (and adult) morbidity and the principal source of mortality in childhood. Unintentional injuries are the leading cause of death and hospitalization among children and teens in the United States, with the highest rates recorded among minority populations. In 2004, nearly 20,000 (44 percent) of the 45,000 deaths of people ages 1-24 were caused by injuries (Bloom and Cohen, 2007:Table 32). The most serious of the nonfatal injuries is traumatic brain injury. The NCS has the capacity to provide critical new information about risk factors (intrapersonal and environmental) and development sequelae of childhood injuries. This measurement focus is responsive to the Children's Health Act of 2000. Each of the three proposed research hypotheses addresses important new areas for injury research that are particularly amenable to longitudinal designs, and each represents a distinct facet of injury research: identification and sequelae of new injury types (rMTBI); longitudinal antecedents of injury-inducing behavior (factors resulting in childhood or adolescent-onset aggression); and long-term sequelae of injury experiences (antecedents and resilience to traumatic life events in childhood).

Assessment: General Concerns About Data Collection

The NCS has thoughtfully addressed some of the measurement issues for the area of injuries. In particular, the research plan effectively discusses the injury threshold issue—for example, the determination of the severity to be counted as an "injury." It proposes a variety of thoughtful data collection methodologies, including initially parent and later child self-reports, health visit logs, and activity diaries to measure the sequelae of injuries that result in the limitation of activities. However, details about the content or implementation of the health visit logs or activity diaries are notably absent—and their implementation in a large national sample could be problematic. Variations in local data collection methods and content could limit the acquisition of national injury data. The relative infrequency of data collection during middle and later childhood years is problematic for surveillance and case identification, especially for less serious injury experiences. No mention is made about using or linking to existing injury databases to further inform the proposed injury data collection efforts.

Assessment: Quality of Hypotheses

The three NCS research hypotheses represent the range and kinds of innovative injury research available from the NCS. However, the quality of the three specific hypotheses is uneven. One of the three proposed injury foci—rMTBI—addresses a newly recognized type of injury and its cumulative adverse effects on subsequent neurocognitive functions. It is unclear, however, if the NCS is appropriately designed or has appropriate data collection capacity to study this specific injury. MTBI is a relatively rare injury, estimated to occur in 7 per 1,000 children visiting the emergency room per year. However, as the NCS research plan notes (Vol. 2, p. A2-280), the multiplicative probability of a repeat emergency room visit among the NCS sample is exceptionally small, with an estimated 100 cases over the entire 20 years of the NCS. Moreover, the NCS provides no direct measure of the mild brain injury itself (via neuroimaging or other means). As proposed, this research hypothesis will be based only on (unspecified) parent reports about each head injury, with perhaps a medical records examination to verify a reported condition. It is also likely that many of these children will not seek medical care. The infrequency of data collection in middle and late childhood years suggests that this research will have to rely on retrospective information. As currently constructed, the NCS does not appear to be an appropriate locus to analyze this hypothesis. Its stated rationale, "the fact that little is known about rMTBIs is a compelling justification for this study," is itself not a sufficient justification.

The second research hypothesis examines the developmental antecedents and pathways of early preadolescent onset compared with adolescent-onset aggressive behavior, with its assumed associated risk for injuries, especially intentional injuries. The NCS could help in disentangling the multiple sources of aggression seen in older adolescents and thereby offer insights into more effective targeted interventions. The current research seems, however, to focus more on the classification of aggression pathways and less on the roles of psychosocial and other interventions (ameliorative or harmful) in affecting the developmental trajectories. Moreover, the links between increased aggression and increased injuries is assumed but would be worthy of study.

The third hypothesis—antecedents and resilience to traumatic life events in childhood—seems almost only metaphorically linked to the injury section of the NCS research plan; perhaps conceptually reflecting on *response* to trauma. The hypothesis focuses primarily on anxiety disorders and the child's response to cumulative exposures to potentially traumatic events—primarily influential life events, but also including physical abuse and injury exposures. While this hypothesis is appropriate for the NCS, it would be much better situated in the child health and development section of the research plan. While the emerging literature on posttraumatic stress disorder (PTSD) underlies this topic, as written the research plan provides little discussion of how injuries (especially intentional injuries) relate to subsequent anxiety disorders, or their relative contribution compared with other life traumas.

Environmental causes of childhood injuries, and clinical responses to injuries, are major issues in discussions of injury policy. Neither is well developed in the research plan. The NCS proposals for the measurement of home and community environments are not well articulated—an issue that is important for other research topics as well (see section on "Exposure Measures" below). Clinical treatment (or nontreatment) of an injury, as a modifier of the injury's long-term sequelae, is never mentioned.

Recommendation

Recommendation 3-6: The NCS should consider replacing research on repeated mild traumatic brain injury (rMTBI) with more nuanced research on other injury-related topics, such as environmental factors in childhood injuries and the effects of clinical response to injury (treatment or nontreatment).

Hormonally Active Agents and Reproductive Development (7)

Description

The NCS proposes to address the effects on the reproductive system from exposure to hormonally active agents (HAAs) in the environment. The study will examine the direct effects of prenatal exposure to phthalates on hypospadias in boys (abnormally placed urinary opening); prenatal exposure to polybrominated diphenyl ether on hypothyroidism, which, in turn, can lead to altered reproductive development; and exposure to bisphenol A on the risks of childhood obesity. The study will also examine the effects of environmental exposures to HAAs on later sexual maturation, including the effects of early childhood exposure to bisphenol A on early onset of puberty in girls; early childhood exposure to phthalates on the development of polycystic ovarian (masculinizing) syndrome (PCOS) in adolescent girls; and exposure to lead in early childhood and at the time of sexual maturation on delayed onset of puberty in girls. The study proposes to examine potential interaction between genetic variations and environmental exposures to HAAs in analyzing the proposed outcomes.

Assessment: Public Health Significance

Exposure to hormonally active agents and their potential effects on reproductive outcomes is an important issue for children's health. Exposure to such agents is widespread, and the potential outcomes are widely observed. Documentation of the associations between HAAs and reproductive outcomes in humans, however, is limited, and speculation about the potential effects is limited to inferences drawn from cross-sectional and retrospective studies or from animal studies. Elucidating the effects of such agents requires a large, longitudinal sample. It should begin before conception in order to investigate the effects of early developmental exposure. The NCS design is therefore well suited for these research questions.

The NCS research plan selects several conditions to illustrate the effects of HAAs: hypospadias, hypothyroidism, obesity, early (and late) onset of puberty in girls, and polycystic ovary syndrome. Cryptorchidism (undescended testes) is mentioned in the section on prevalence/incidence, although it is not clear whether this is meant in isolation or in conjunction with hypospadias. Specific environmental exposures to be examined include polybrominated diphenyl ether, phthalates, bisphenol A, and lead. The focus on these outcomes and exposures represents a strength of the NCS.

In addition to the chemicals listed above, there are many other hormonally active compounds with widespread human exposure. They include, among others, UV sunscreens and parabens (used as a preservative in foods and cosmetics). An archive sample of urine and blood could be collected for future analysis of these classes of compounds. In addition, information could be collected on dietary consumption of phytoestrogens during pregnancy. A urine archive could also be used to measure phytoestrogens.

Assessment: Methodological Concerns

For none of the conditions identified does the NCS research plan provide a clear definition of the outcome. Hypospadias and PCOS can vary significantly in severity, and it is not clear that all degrees of severity are equally important. PCOS may also be difficult to detect in adolescents and some ethnic groups (Norman et al., 2007); it may not become evident until childbearing is attempted. There is controversy about which definition of PCOS to use and which diagnostic procedures to implement (Norman et al., 2007; Toledano and Nelson, 2007). Moreover, other causes of hyperandrogenism must be eliminated, and this diagnostic issue is not raised in the research plan. The issue of premature puberty in girls and changing secular trends remains controversial, may differ for different ethnic groups, may not actually be associated with changes in the onset of menses in girls, and has primarily been associated with obesity (Himes, 2006). Thus, the exact targets of analysis are unclear.

Missing from the list of outcomes is the assessment of pubertal onset in boys, which can be easily evaluated and should be included in the research plan. At a minimum, Tanner staging should be used to assess pubertal developments in both boys and girls (Tanner, 1962). The research plan indicates that the NCS will refine its approach to ascertaining the onset of puberty, and there is time do so. It is not clear from the research plan if children will have a physical examination as part of the study. However, for defects present at birth or early childhood, there is not that luxury of time.

Of relevance for statistical power considerations are prevalence estimates for the specified outcomes, which range from 0.8 percent for hypospadias, through 6 percent for PCOS (women of all reproductive age ranges, not just adolescents), to as high as 52 percent for 7-year-old and for 8-year-old African American girls showing signs of precocious (early onset) puberty (Herman-Giddens et al., 1997). No prevalence is given for hypothyroidism. For most outcomes, the NCS research plan notes that only half the sample can be used (either male or female), thus reducing power, and prevalence may be lower depending on the ethnic group. The expected NCS sample sizes in the research plan for estimating power are derived from studies with relatively high levels of exposure. The plan argues that the power should be sufficient to detect differences if the prevalence of exposure is higher than the studies in the literature. However, this remains an assumption.

While recognizing that there is widespread general population exposure to most of the hormonally active compounds of interest, it is not clear that the NCS sample design will result in an optimal distribution of exposures or outcomes. The reason is that relatively high exposures may be needed to detect associations (Toledano and Nelson, 2007). Specifying one or more substudies that enriched the sample with high-exposure population groups could be valuable.

Recommendation

Recommendation 3-7: The NCS should develop refined and detailed protocols for studying reproductive development outcomes, which, as presented in the research plan, often lack clarity in measurement and research design. Outcomes that are measured at birth for which there is little time to refine research protocols require immediate attention. The NCS should use results from the Vanguard Centers, such as estimates of the prevalence of specific

reproductive development outcomes, to assist in protocol development, and it should consider the usefulness of substudies of high-exposure population groups.

EXPOSURE MEASURES

In this section, we provide assessments of the proposed measures of exposures. These measures include the demographic and socioeconomic characteristics of parents, the chemical environment (persistent chemicals, nonpersistent organic compounds, air pollutants), the physical environment (housing and neighborhood conditions), psychosocial exposures, biological exposures, and genetic markers. (See NCS Research Plan, Vol. 2, App. F-1, for details of planned interviews and other methods to collect various kinds of exposure data from prebirth through age 24 months.)[1] We then note important types of exposure data that are missing and should be added to the NCS research plan when resources permit. We conclude with a discussion of the desirability of facilitating linkages of the NCS to other data sources that could enrich data on exposures.

Demographic and Socioeconomic Measures

The NCS properly recognizes the need to collect a variety of interview-based demographic data for participating children and their parents, including age, gender, race, and ethnicity, together with measures of socioeconomic status (SES), such as education, occupation, home ownership, income, and family structure, for the parents. These data are needed as a means to classify population groups of interest, as controls for the confounding influences of factors associated with demographic and social position in causal analyses, and as dimensions for the focus on health and developmental disparities that is mandated by the Children's Health Act of 2000.

By and large, the research plan lays out a cogent strategy for gathering these kinds of data, relying on widely used instruments from the decennial census, the American Community Survey (ACS), and the Survey of Income and Program Participation (SIPP). Our discussion of these measures focuses on potential problem areas.

Education, Employment, and Income

The research plan calls for measurement of educational level, employment status, and income of the mother and father at the initial face-to-face

[1] We note that the information in Appendix F-1 does not always agree with information in other parts of the research plan.

visit using standard questions from SIPP, Census 2000, and the ACS, with updating at the in-person visits during pregnancy and following the birth of the child. With regard to education, we applaud the recognition that parents, particularly younger parents, often acquire more education after the birth of children, and that the NCS will need to ask about completed schooling periodically over the course of the study.

Income and poverty status are key constructs of socioeconomic status, and, given their volatility, it is important to measure both recurrently. The research plan details how this would be done between prepregnancy and age 1 but not in the 24-month phone interview. The proposed SIPP income question battery may be too detailed to be worth the interviewing time; the Current Population Survey (CPS) or other surveys (for example, the Panel Study of Income Dynamics) provide validated shorter alternatives.

It is unclear what "Supported by income" in the measures grid means. Money mingles within households, and it is sufficient to get a good accounting of household-level income and composition.

Maternal employment will obviously vary a great deal surrounding the birth of the child, with relative high employment rates during the early stages of pregnancy, near-zero rates around the birth, and then increasing rates during the months following the birth. A key design issue is whether measures of employment (and income) will be gathered as "time of interview" variables or in an event-history format (for example, month by month) that would provide a "motion picture" of parental employment across the survey period.

Given the constraints on interviewing time, it is probably not worthwhile to invest in a detailed 21-year motion picture of all aspects of maternal employment. But it may well be worthwhile to invest interview time in a month-by-month employment history checklist (for example, whether employed or not), beginning with the month prior to the enrollment interview and extending (through retrospective questions in subsequent waves) at least until the child's first birthday and perhaps throughout childhood. More details about month-by-month employment (for example, wages, work hours) may well require too much interviewing time relative to their analytic value, but wages and work hours questions modeled after the CPS should probably be asked about the week prior to each interview.

In addition to employment status, work hours, and wages, the research plan should include occupation. It is not clear whether and how the NCS intends to generate data on occupation, which is an important indicator of socioeconomic status. Questions about occupation should be asked (for example, using questions from the CPS) at the prepregnancy interview and then again at, say, the child's first birthday as well as in later interviews.

Immigrant Status

It is not clear what immigration measures are planned, other than a standard "country of origin" question. Given health differences between immigrant and nonimmigrant children, the cultural implications of country of origin, and the difficulties of accessing health care for undocumented immigrants, it is vital to ascertain immigrant generation, languages spoken (primary, secondary), acculturation, and, if possible, the legal status of parents.

Recommendation

The NCS measures of demographic and socioeconomic variables are generally well planned. The highest priority is to include more information relative to immigration status.

> **Recommendation 3-8: The NCS should add to its well-planned battery of demographic and socioeconomic measures questions on immigrant generation, languages spoken, and, if possible, the legal status of the parents and child.**

Chemical Exposure Measures

As described in the research plan, the primary purpose of the NCS exposure assessment is to enable the epidemiological analysis of associations between priority exposures and priority outcomes that are described in the study hypotheses. The study will archive environmental chemical (and biological) samples for potential future analysis of other exposures that are not deemed a priority in the current proposal. Broad classes of environmental chemical exposures include, among others, metals, modern chemicals, pesticides, air pollution, and drinking water disinfection by-products.

General Comments

The overall goals of exposure assessment strategies in epidemiological studies are to identify populations with a wide range of exposures to maximize power and to identify populations with high exposure levels, because associations are more likely to be detected in such a setting. In addition, measurements need to be accurate and precise. It is well recognized that in environmental epidemiological studies, exposure assessment is expensive and there are numerous limitations.

One concern, which applies broadly for the NCS, is whether the national equal probability sampling strategy will enroll sufficient numbers of children living in different conditions across the United States to ensure a range of exposures to environmental chemicals. For some contaminants, such as air pollution and pesticides, there are segments of the population with high exposures, and the NCS could usefully consider ways to enrich the study population with these segments through substudies or adjunct studies. Doing so would ensure that the NCS has large enough samples of children with the conditions of concern for the main hypotheses, as well as large enough samples to understand the causes and consequences of health disparities. The NCS could also usefully consider a plan to reassess the sampling scheme as recruitment progresses in relation to the exposure variability among participants (see Chapter 4).

The NCS exposure assessment strategies, as described in the research plan, strive to measure the magnitude, frequency, and duration of exposure. The study should also strive to measure exposures at the relevant etiological exposure window, which generally requires repeated measurements for later sophisticated statistical analysis. The research plan makes some general statements on the matter, but detailed plans and the rationale for the timing of data collection need to be provided to ensure that this critical—and difficult to accomplish—aspect of the exposure assessment strategy is well specified.

The research plan mentions the need for a validation study of exposure measurement in subsamples of the cohort, but it does not describe how this will be carried out and incorporated in the overall analysis. The plan should provide further details, which will of course vary depending on the characteristics of the exposure of interest.

Figure 9-1 in the research plan (NCS Research Plan, Vol. 1, p. 9-4) is used to illustrate that questionnaire data, observations, environmental measures, and biological measures will be combined to assess exposures. However, the figure alone is not detailed enough to provide a description of how this will be accomplished. The plan needs to specify how data from these multiple sources will be combined into exposure metrics for the priority candidate exposures outlined in the proposal.

Finally, the NCS should give consideration to collecting biological samples for archives that can be later analyzed for hormonal activity. Bioassays exist to measure total estrogenicity or antiandrogenicity of a biological sample. This approach is less expensive and potentially more relevant biologically than standard chemical analysis for many compounds.

Persistent Chemicals

The research plan states that the assessment of maternal persistent organic chemical (POC) burden, which is an indirect measure of fetal exposure, can be obtained from maternal blood taken before or during pregnancy or obtained from maternal blood, milk, or adipose tissue taken soon after parturition. Until recently, this approach seemed reasonable for POCs. However, Bloom et al. (2007) showed that POC levels vary across pregnancy, which indicates that the timing of the POC sample collection should be reconsidered.

The research plan further mentions that POCs will be measured in urine, but urine is traditionally not used as a medium to measure POCs. The study needs to make clear how such measurements will be conducted.

The statement on page 9-4 of the plan (NCS Research Plan, Vol. 1) that "the purpose of testing at multiple time points is to determine when the exposure(s) occurred" is not entirely relevant for POCs. It is difficult to determine when exposure occurred for POCs unless there is a single acute high-exposure source. The paragraph goes on to state that dust samples will be collected from homes to help distinguish between past high-level exposures and recent lower level exposures, but it is not clear how the use of dust samples will accomplish this. Furthermore, it is likely for POCs that there are low-level, continuous exposures rather than past instances of high-level exposures. The research plan needs to provide further specificity on the assessment of POC exposure, including the use of dust samples and the timing of collection.

In Table 9-1 (NCS Research Plan, Vol. 1, p. 9-5), modern chemicals, such as parabens and UV sunscreens, need to be added to the list for analysis of archival samples. These chemical classes may alter endocrine signaling.

Assessing the use of clothing treated with flame-retardant chemicals by questionnaire, as intended, is likely to be difficult, since parents may not know which clothing articles have been treated. In addition, there are dozens of flame retardants, and it is therefore difficult to determine which specific ones are used in specific articles of clothing. It would be useful if there were pictures or product names to help the parents identify the relevant articles of clothing. Given the high prevalence of exposure to brominated flame retardants among the U.S. population, identifying sources of exposure is a worthwhile aim of the NCS.

Nonpersistent Organic Compounds

The NCS plans to measure volatile organic chemicals (VOCs) from samples of soil, water, dust, and air. Water collections are to be made at homes served by a community water supply. It is not clear whether samples will be taken from homes served by wells. If not, there may be missing data related to potential sources of exposures. For example, well water is more likely to be from rural homes with different exposure patterns for pesticides compared with city dwellings.

For semivolatile organic chemicals and nonpersistent nonvolatile organic chemicals, the research plan recognizes that these compounds are rapidly metabolized and therefore multiple measurements are needed. However, the specific timing of sample collection, the number of samples collected, and the statistical methods used to analyze these results need to be clarified. Furthermore, it is not clear how these multiple (repeated) exposure data points will be incorporated into the study and used in the testing of the priority hypotheses.

The NCS presents a unique opportunity to improve understanding of the use of biological markers to estimate historical exposure for nonpersistent chemicals. For example, much concern exists regarding the potential health effects of organophosphate pesticides and bisphenol A, chemicals with short half-lives. However, there is limited information about whether prior exposure to such toxicants can be estimated by identifying signatures using "omics"[2] or through the development of novel technological exposure assessment approaches. This opportunity deserves consideration in the design and execution of the NCS.

For nonpersistent pesticide analysis, to minimize failure to detect exposures, the NCS plans to collect environmental samples only when a recent pesticide application was reported or in agricultural areas. It is not clear how "recent application" will be defined (e.g., days, weeks). The definition of "recent" should be partially dependent on the specific pesticide used, since pesticide half-lives vary in the home environment. The NCS notes that water and soil samples for pesticides will be collected only in rural areas, yet there are data showing high levels of indoor pesticides in urban areas, specifically in public housing and apartment complexes due to the use of insecticides for pest control.

The research plan states (NCS Research Plan, Vol. 1, p. 9-7) that the parent will answer questions on the use of teething rings and pacifiers by the child. It will be important to collect product names, since the presence and percentage of chemicals, such as phthalates, present in these products may vary widely.

Air Pollutants

The research plan states that personal sampling, using a portable monitor, detector tube, or other technology, is burdensome, and therefore indoor air sampling will be performed to measure exposures to various pollutants.

[2]Omics is a general term for a broad set of disciplines for measuring and analyzing all components interactions from genes to proteins to metabolites.

However, personal sampling data are usually superior to area sampling data for assessing individual exposures. Statistical methods could be used to relate indoor and personal sampling results from a subsample of subjects, and the results could be applied to the larger cohort. The NCS plans to collect particulate matter less than 10 microns. Currently, there is much more interest in particulate matter less than 2.5 microns and in ultrafine particulate matter, because these smaller particles are more strongly associated with risk in air pollution studies, so air samples collected in the NCS need to allow for a 2.5 micron and ultrafine cut point measurement.

The research plan states that biomonitoring has a limited role for assessing exposure to air pollutants. Yet urine samples are often used to measure exposure to some of the metals found in ambient particles, and an aliquot of urine could be archived for potential measurement of metals relevant to ambient air pollution exposure.

More generally, the NCS needs to reconsider its plan not to collect biological samples for measuring VOCs and other pollutants, even though most media (expired air and urine) are easy to collect and future analytical needs may outweigh the time for collection. Granted such biospecimens have short half-lives, but they are relevant for assessing personal exposure.

The NCS should consider incorporating the collection of novel biological media that may be of potential value for exposure assessment. These data include meconium and the shedding of primary teeth. The former has been demonstrated to be of potential value for measuring organophosphate exposure, and the latter has been well demonstrated to be of significant value for estimating cumulative exposure to lead in children.

Finally, there are few details or protocols for data and biological sample collection from fathers. However, there is a growing literature identifying paternal exposures as of potential importance to offspring, possibly through epigenetic mechanisms (e.g., the impact of paternal age on the risk of offspring schizophrenia). It is recommended that the NCS consider collection of biological samples and data from fathers, when appropriate.

Other Concerns

Recent experimental evidence suggests that there are transgenerational effects of some modern chemicals, possibly through epigenetic reprogramming. Such considerations raise the question as to whether the full effects of some chemical exposures can be detected in the proposed time frame of the NCS. The development of more than a two-generation cohort (parent-child) is beyond the scope of the NCS, but investigators in the NCS could consider the possibility of following subsamples of participating children into their adult years so that a multigenerational study would be feasible (for example, mother-child-grandchild), if results from the current study warrant such assessments.

Projected response rates for the NCS are only 75 percent at the stage of enrollment. Because nonparticipation in the study might be related to characteristics that predict environmental chemical exposures, the NCS needs to develop methods to measure such relationships and their effects on analysis and to make appropriate adjustments to the data (see Chapter 4).

There are important ethical and methodological issues involved in the issue of whether and how to report environmental chemical exposure results to study participants (see Chapter 5). Some exposure measures are currently not interpretable as to the level of potential risk. For some exposures, it is not even clear what is considered high or low exposure. Other exposures, such as to high lead or mercury levels, are known to carry increased risks of adverse health effects. If study participants are told of such results, one hopes that they would alter their lifestyle, home environment, or diet to reduce exposure and thus potential risk. The NCS needs to develop detailed plans for reporting results to participants and for taking account of any consequent changes in behavior that reduce exposures.

Recommendations

Our discussion has identified a range of issues for the NCS program staff to consider. The most important area to address concerns the potential of biological samples for assessment of chemical exposures.

Recommendation 3-9: The NCS should consider the use of personal air sampling methods for a subsample of participating women and their children for measuring exposure to air pollutants.

Recommendation 3-10: The NCS should incorporate methodology to measure paternal exposure to environmental chemicals (both persistent and nonpersistent). More generally, the NCS should consider collecting for fathers, not only chemical exposures, but also biological samples and interview data on paternal characteristics that may affect children's health and development to the same degree as it collects such information for mothers.

The collection of biological samples from even a subsample of fathers would enrich the NCS study aims and provide the opportunity to assess paternal contributions to children's health.

Physical Exposure Measures

The NCS research plan recognizes the influence of the physical environment, in the form of housing and neighborhood characteristics and conditions, on child risks and outcomes as children age. Early measurement properly focuses heavily on the home environment, given that preschool-age children spend the majority of their time in the home. Key housing dimensions include the physical configuration and condition of the residential structure, which would be measured with a combination of respondent questions and interviewer observation. Other measures to be collected include household appliances, such as a vacuum cleaner, clothes dryer, and vaporizer (NCS Research Plan, Vol. 2, App. H, p. H-1). Community measures include visual assessments of the neighborhood in terms of type and condition of housing and presence of industries, businesses, recreational areas, traffic, and graffiti; geographic information systems data on traffic, nearest hospitals, recreational facilities, and Superfund and brownfield sites; aerial photography of topographical features; and census population density.

Methodological Concerns

Currently, the NCS research plan fails to acknowledge any critical or sensitive periods during which children's health and development may be most vulnerable to exposure to specific housing and community factors. Hypotheses regarding such periods would provide opportunities to test for their presence and effects. Given that data collection burden is appropriately emphasized as a concern, the specific measures being proposed should be linked to the developmental stage of the child rather than collecting all measures for all children of a specified age. Greater precision in the data-gathering would allow more in-depth measures to be collected at developmentally appropriate assessment points and may eliminate the need for other measures.

The indicators and the frequency at which they are measured appear not to capture the dynamic nature of some of the home or community exposures. The research plan appropriately notes that many structural features do not change frequently (provided families do not move), yet other exposures may vary daily or weekly, and capturing information at several points over a short period of time versus at a single point in time is state of the art but is not mentioned in the plan. A stronger rationale for some of the listed factors (for example, location of the garage, outdoor additions) should be provided, as their relevance is questionable.

Given the wide variety of housing and community conditions to be encountered, it is not clear whether the existing measures being proposed for use will be valid across types of settings and housing. No mention is made of using the Vanguard Centers to further validate or refine these measures.

It is not clear how physical exposures are going to be modeled in the statistical analyses. For example, how will data from aerial photography on rivers and mountains be used to create indicators of exposure? The links between sources of data on physical exposures, creation of indicators of exposure, and the hypotheses for child outcomes need considerable strengthening.

While turning to readily accessible information as a source of data on community characteristics is a cost-effective strategy, careful attention should be given to how such data are used to construct indicators of neighborhood features. Issues to be resolved include using single factors versus composite indices, creating concurrent versus dynamic historical profiles of the neighborhood, and choosing the appropriate unit of analysis at the community level (for example, census tracts, administrative neighborhoods, etc.). Moreover, a good definition of neighborhood should be provided, given the interest in various types of exposures at the community level. Each of these issues should be considered in light of reformulated and refined hypotheses about the role of community factors in shaping risks and adverse outcomes.

Recommendation

> **Recommendation 3-11: The NCS should provide a clearer rationale for some of the housing and neighborhood conditions it proposes to measure and revisit its data collection plans to ensure that needed measures are obtained at developmental stages when children may be more vulnerable to risk factors. The goal should be a set of measures and data collection plans that are optimal with regard to analytic utility and response burden.**

Psychosocial Exposure Measures

The NCS research plan proposes to collect information on an exceedingly diverse set of psychosocial exposure measures, including family process measures, such as the home environment and domestic violence; maternal mental health measures, such as depression; psychosocial stress measures, such as perceived chronic stress; social support; health behaviors, such as smoking; and features of child care. At the same time the research plan reflects a relative lack of attention to fathers, family structures and transitions among structures, and other significant support figures. The latter may be particularly important in enhancing resilience in the face of adverse family characteristics. Each of these domains has a robust research literature providing justification for literally hundreds of possible measures of psychosocial influences on child health and development. The NCS cannot possibly cover even a fraction of these possible subdomains, and rarely can it afford to allocate enough interviewing or observation time to collect the highest quality measures in any of them.

Overall, however, the research plan does not provide a clear justification for its implicit set of decisions across the daunting set of psychosocial measures to collect. It is clearly not appropriate for reviewers of the research plan to merely suggest that the NCS expand to collect measures of neglected subdomains, or to devote considerably more interviewing time to obtain gold standard versions of the included domains. The existing plan for data collection already imposes heavy burdens on respondent time and attention. Rather we call for a careful elucidation of the choices made with specific reference to goals, hypotheses, and relevant outcome measures.

Relating Exposure Measures to Outcome Measures

The panel thinks that the best way to make decisions about what psychosocial exposure methods to collect rests on a refinement and focusing of the domains of interest in terms of outcomes. For example, if conduct disorders were to become a major focus, one would want to assess disciplinary interactions at key points through the life span (see, e.g., Patterson and Forgatch, 1995). In addition, one would want to tackle the tricky problem of obtaining reliable information on family criminal histories and violence in the home and, beginning at least in middle school, detailed information on peer associates. Decisions about where to invest respondent time and study dollars need to be made in light of decisions about what kinds of variation and what problem outcomes the study wishes to focus on.

Obtaining High-Quality Measures

A general observation about the psychosocial measures concerns the frequency and type of planned data collection. It is striking that there are no plans for an in-home interview between age 1 and at least age 5. This gap precludes accurate measurement of such key elements of the psychosocial environment as the home learning environment during a key phase of development. To obtain high-quality assessments, the NCS needs to conduct more frequent home-based interviews, performed by trained interviewers, at least for a subsample (see Chapter 2).

Given its time constraints, we support the choice of the relatively abbreviated Infant/Toddler Home Observation for Measurement of the Environment (HOME) scales for assessing the physical, emotional, and learning environment. We question, however, that the study plans to invest in a videotaped assessment of mother-infant interaction at 6 months, rather than a high-quality assessment of attachment at 12 or 18 months, such as the Strange Situation, which is feasible to administer (Ainsworth and Bell, 1970).

Finally, we note that, as part of its contribution to the development of state-of-the-art methods, the NCS could engage in the development of high-quality but quick-to-administer measures of psychosocial exposures. The psychological, social, and behavioral sciences have developed reliable, valid, and theoretically informed measures for assessing psychosocial exposures and outcomes. However, at the present time, many of these assessments are lengthy and costly to administer. They demand coding of videotaped interaction, use of laboratory-based equipment, or the like. Advances in knowledge frequently depend on addressing these limitations. For example, a MacArthur Network developed a parental assessment of language development that has led to an explosion of knowledge (compared with transcription of audiotape), and the Early Childhood Longitudinal Study developed a short form of the Wechsler Preschool and Primary Scale of Intelligence (WPPSI). Another successful example is the HOME scales for assessing the physical, emotional, and learning environment, which the research plan includes. The NCS, with its long frame of observation (over 21 years) and large sample, is well suited for this type of methodological research conducted through substudies and adjunct studies. The NCS, in turn, can benefit from the development of high-quality, easily administered assessments that can be used to replicate its findings in other studies.

Recommendations

Recommendation 3-12: The NCS should reconsider its psychosocial measures to ensure that they will provide high-quality data for outcomes of interest for child health and development. In the face of resource and respondent burden constraints, the NCS should emphasize the quality and analytic utility of information, even if some measures must be dropped in order to substitute other assessments more desirable on various grounds.

Recommendation 3-13: The NCS should dedicate a portion of funds to support research and development of reliable and valid instruments of key psychosocial measures that are practical and economical to administer.

Biological Exposure Measures

The NCS investigators plan to obtain measures of biological exposures at several points in time. The measures fall into six main categories: allergens, markers of infection/inflammation, maternal glucose metabolism, endocrine markers, parental medical history, and other health behaviors. Most are to be obtained through a combination of questionnaire and direct measurement.

The panel of allergens includes those related to cats, dogs, mice, rats, cockroaches, and mites, as well as a panel of molds and pollen to be obtained from regional ambient monitoring. Infections are generally to be obtained by a medical history provided by the parent with ascertainment of a variety of inflammatory markers. Maternal glucose metabolism during pregnancy is to be obtained from studies done as part of clinical care and obtaining a hemoglobin $A1_c$, a measure of the average glucose for several weeks. The endocrine markers of interest include maternal thyroid function and cortisol as a measure of stress. Cortisol measures will be obtained from the mother and child on several occasions, and exposure to stress will also be ascertained by questionnaire. Parental history will include attention to the presence of chronic illness, diet, physical activity, use of tobacco and alcohol, use of illicit drugs, and use of medications and supplements. Other health habits to be ascertained include dental health (by questionnaire), maternal sleep patterns, and maternal douching.

This list of potential markers of biological exposure is comprehensive. Several, such as diet, allergens, and maternal physical activity, are to be obtained by well-standardized methods. The use of standard and well-tested approaches to obtaining data on these biological exposures is a strength of the study.

Methodological Concerns

Less clear is whether plans to obtain other exposure measures are as state of the art. Information on child infections is to be obtained primarily by maternal history. The recall period for this information (6 months in the first year) is greater than that generally considered desirable to produce accurate data on relatively brief illnesses. Moreover, it is unclear that mothers will be able to report accurately about the type of infecting organism, information that could prove vital to analysis of subsequent growth and development. The timing of the measurement of inflammation relative to the infection that may have caused it is also not clear, raising questions about what types of infection/inflammation may be missed.

Interest in maternal glucose metabolism in pregnancy is driven primarily by its potential effect on birth defects and child obesity. However, the planned measure of glucose metabolism during pregnancy will be an average measure over some weeks that is not obtained close to the onset of pregnancy, when organogenesis is occurring. The research plan indicates that the study did not think more active measures of glucose metabolism could be obtained given that the first data collection point during pregnancy is a home visit, but it does not speculate on the effect that decision might have on the testing of core hypotheses.

Likewise, the research plan notes that the only endocrine measures of interest are maternal thyroid function during pregnancy and cortisol levels in the mother and infant at various times. With regard to the former, it should be noted that child hypothyroidism may be a mechanism for the linkage between hormonally active agents and reproductive outcomes, so it is not clear why only maternal thyroid function is of interest. It would have been helpful in the discussion of cortisol levels to have some sense of the time over which cortisol will represent an accurate marker of stress in order to assess whether the frequency of data collection will be adequate.

The parental medical history seems quite comprehensive. In conjunction with measures of maternal physical activity, there are plans to estimate infant physical activity at 6 and 12 months through developmental observations and questions about usual activities. It is not clear what significance infant physical activity at this age has, whether developmental observations can actually characterize infant activity, or whether questions about usual activities provide information that is comparable to concerns about parental activity.

Other health behaviors are to be elicited by questionnaire. Ascertaining intensity and frequency of intake of alcohol, and even more so of illegal drugs, from interviews with pregnant women is potentially problematic. In particular, exposures are likely to be underestimated. It is troubling to see little if any discussion of how this potential threat will be addressed. Simply promising anonymity may not suffice. One plan that seems particularly problematic is ascertaining dental health through questionnaire and not examination. With the poor coverage of dental health care in the United States, it is likely that many dental problems will go unrecognized by the parent. To the extent that dental disease, such as peridontitis, is a risk factor for prematurity and other child outcomes, inaccurate data will make it more difficult to establish the link.

A serious concern is that there does not appear to be a conceptual framework guiding the selection of the various biological markers of exposure. In particular, it is not clear that all aspects of parental health that might be relevant to people's ability to be effective parents will be ascertained.

Recommendation

Most important among the concerns we have raised about biological exposures is the timing of data collection.

> **Recommendation 3-14: The NCS should review some of the proposed measures of biological exposures, such as maternal glucose metabolism and child cortisol levels, to ensure that the proposed times for data collection are appropriate for capturing the underlying exposure.**

Genetic Measures

Each project outlined in the NCS research plan includes investigation of how genetic variation contributes to variation in risk of the study's key outcomes, ranging from childhood obesity to neurobehavioral traits. With advances in high-throughput genotyping technologies, it is now possible to directly measure hundreds of differences in particular genes and millions of mutations in the whole genome of large numbers of individuals to elucidate the genetic contributions to a human trait or disease. In general, human variability in any trait arises from a complex interaction among genetic variations and environmental variations. The association studies proposed by the NCS are currently the most efficient strategy to explore the putative contribution of genetic variations or gene-environment interactions to variations in disease risk.

The basic design and analysis principles of genetic association studies have been well established for decades, albeit with continual evolution in such areas as study designs (e.g., family-based association studies, case-only designs); genotyping (e.g., multiplexing assays, array-based genotyping); dealing with underlying genetic confounders, such as variations in population groups studied (Haines and Pericak-Vance, 1998); and reducing the probability of false positive results (e.g., Benjamini and Hochberg, 1995). All of these issues appear to be adequately addressed in the research plan. There are, however, several major weaknesses in the plan proposal that should be addressed before the genetic component of the project is actualized.

Changing Science and Scientific Standards

The field of genetics and genomics is changing rapidly. Just within the past year, dozens of genome-wide association studies have identified new genes and variations that are involved in such complex traits as blood glucose levels, obesity, height, and variation in common disease risk (English and Butte, 2007; Fox et al., 2007; Hayes et al., 2007; Peeters et al., 2007). In many of the NCS proposed analyses, in contrast, the old approach of studying "established" candidate genes (i.e., genes that are thought to be involved in a disease or trait because of the knowledge of underlying biological pathways) is put forward and is a major weakness of the proposed genetic studies.

The NCS research plan does discuss in several places the use of new technologies, namely, gene expression profiling and epigenetic profiling. These new methods may be of some use in unraveling the potential molecular mechanisms underlying genetic associations. However, in the NCS, these biological signatures will be measured on tissues—namely, components of blood—that may not be relevant to the trait or disease being investigated. For example, neurodevelopmental outcomes are not likely to be associated with transcriptomic variation in lymphocytes, which will be the only biological tissue available for study. Great care must be exercised in making inferences from these transcriptomic and epigenomic types of studies because they are, in many cases, studies of convenience rather than studies designed for their scientific rigor. Although it may be convenient to measure gene expression or epigenetic changes in blood samples, there is very little evidence that the gene expression profiles in this tissue are biologically relevant to the neurobehavioral outcomes or other outcomes investigated in the NCS.

The scientific standards for genetic association studies are also quickly changing, as scientists come to grips with the limited success of the past two decades of genetics research, which has failed to identify the key genetic factors with reproducible or replicable effects on common human diseases or traits. The previous lack of high scientific standards for publication has resulted in the dissemination of false information (including false positive results from genome-wide association studies), the waste of millions of taxpayer dollars, and an increase in genetic deterministic thinking among the public. Three systematic studies of the genetic association literature have documented the extent of the problem (Hirschhorn et al., 1999; Ioannidis et al., 2001; Lohmueller et al., 2003), in which the odds of a published genetic association being replicated when 3-4 other studies were conducted were approximately 1 in 30. Moreover, Lohmueller et al. (2003) found that less than half of what are considered established susceptibility genetic markers pass the standards of a rigorous statistical meta-analysis. Cited reasons for the lack of reproducibility include genetic and environmental differences across the populations being studied, low statistical power, and misclassification of disease outcomes (e.g., Cardon and Bell, 2001; Colhoun et al., 2003; Freedman, Reich, Penney, et al., 2004).

In the current NCS research plan, the strategy for investigating genetic associations appears to lack an appreciation of the more rigorous standards now being imposed by the scientific community. There are multiple reasons for adopting a high standard in the NCS by which genetic association studies must be internally and, optimally, externally validated before any type of publication or media release. First, the field of genetics research is finally imposing its own higher standards. Second, given the powerful implications of genetic information (for example, stigmatization and discrimination) for children if a genetic marker of a trait, say ADHD, is identified in the NCS, there must be a mechanism in place for validation at both the population and molecular levels to avoid the reporting of false results, many of which have already flooded the literature and mass media from other sources.

Gene-Environment Interactions

A major strength of the NCS research plan is its emphasis on gene-environment interactions. However, the implications of this emphasis for the measurement of exposures are not fully appreciated. The lack of adequate measures of psychosocial and behavioral variables is particularly important, for it severely limits the ability to examine gene-environment interactions that are likely to affect obesity, neurobehavioral phenotypes, asthma, and pregnancy outcomes (Institute of Medicine, 2006).

The genetic measures proposed in most analyses are quite limited. For each of the major outcome areas, genetic associations are proposed based on previous candidate gene association studies, many of which would not pass more modern standards of evidence. Furthermore, the variation within these genes is not well represented or discussed in the research plan. Most of the genetic variations targeted for study are not considered causative, have no known biological function, and are simply markers of the effect of a currently unknown genetic variation. Thus, more variations within each gene need to be considered if the true goal is to understand the genes responsible for modifying the effects of environmental exposures.

Another weakness in the research plan is the lack of detail on how genetic variation will be measured. There are considerable cost considerations, which are not dealt with adequately. For example, it can be as expensive to genotype 20 single nucleotide polymorphisms (one at a time) as it is to genotype thousands (or millions) of them with array-based genotyping platforms. Given the size and scope of the NCS, there is also the need to consider gene-environment correlations in analyses, in addition to gene-environment interactions.

Data Collection Concerns

Because the DNA sequence variation proposed to be investigated in the NCS genetic association studies does not change over time, the one-time collection of a blood sample for genetic analysis (summarized in NCS Research Plan, Vol. 2, App. G) will be adequate for the genetic analyses proposed. However, unlike DNA variation, gene expression and epigenetic variation will dynamically shift over the life course of the mother, father, and child. They are also tissue-specific. Currently, there is no proposed collection of blood samples for studies of gene expression profiling of parents. In addition, only a single cord blood sample from the child is being set aside for transcriptomic and epigenomic studies. This is a major oversight and an inconsistency between the neurodevelopment analysis plan (NCS Research Plan, Vol. 2, App. A-2) and the biospecimen collection plan (App. G). Although this oversight can be easily remedied by collecting and storing additional blood samples for gene expression studies at other time periods, it must be actively addressed or this opportunity will be missed. As we cautioned above, the NCS must be aware that examining gene expression and epigenetic patterns in blood may not be relevant or appropriate to study with such outcomes as the neurodevelopment outcomes. Great care must be taken, since false positives are much more likely than true positives in these domains.

Appropriateness of the NCS for Genetics Research

The greatest potential benefit of the NCS to the field of genetics is its linkage between prospective environmental exposure data and high-quality, high-density genetics data. The field of gene-environment interaction has very few examples in which prospective environmental exposure data are linked to genetic susceptibility data. Virtually all genetic association study designs are either family-based or case-control designs, because DNA sequence is static and can be measured at any time. The most difficult aspect of a genetics study is finding probands and conducting sound phenotyping. Since it is assumed that genotype precedes phenotype, genetic studies nearly always focus on phenotyping first and genotyping second.

The value of the NCS is that it can solve both problems—that is, some genetic studies can be conducted prospectively in this NCS cohort, and environmental studies on the same disease will have good unbiased, exposure data. This is a unique opportunity for high-quality gene-environment interaction research. After 22 years, there will be sufficient cases for several diseases (autism, ADHD, asthma) to conduct nested case-control GXE interaction studies with prospective environmental data. This is of critical value. The concept of critical developmental windows indicates that the timing of exposure matters a great deal. In either case-control or family-based association designs, sampling is dependent on case designation. With either design, environmental exposure is nearly always either measured cross-sectionally or retrospectively. Exposures during developmental windows cannot be reconstructed retrospectively. Case-control designs for gene-environment interaction ignore what is known about gene expression changes in development. Genes are not always active throughout each life stage or may be more or less active during different life stages (e.g., hormones and adolescence). Logically, a chemical or other environmental exposure that modifies a genetic variant may only do so when the exposure corresponds with a developmental stage. Unless the exposure dose *and* the timing of exposure are matched to the genetic variant, critically important gene-environment interactions may be missed. This is the greatest value of the NCS: Not only will it provide measures of environmental exposures of interest, but it will also provide their timing and relationship to genetic variation. This is not well articulated in the research plan, yet it is the study's greatest strength in this field.

Recommendations

Recommendation 3-15: The NCS should adopt a clear mechanism by which genetic association studies are internally and, optimally, externally validated before any results are published or released to the media. The NCS should also revise its proposed "established" candidate gene approach to take advantage of the new information emanating from the current wave of genome-wide association studies, with appropriate replication.

Recommendation 3-16: The NCS should consider consolidating its genetic studies in order to reduce costs and to coordinate the best science at the least cost to the project. One approach would be to simply collect the biological samples and properly store them for later genetic analysis when a better selection of polymorphisms and cost-effective genotyping across studies are possible.

Missing Exposures

Access to and Quality of Services

Clearly, even such a large study as the NCS is limited in the data that can be collected. A discussion of what exposures were considered and rejected and why would have been useful for the panel in its review. In particular, a notable omission is information on access to services, especially health services, both as potential mediators of outcomes and as factors in the accuracy of information obtained through maternal report.

A substantial amount of information—such as diagnoses of child health problems—is to be ascertained through interviews with the mother, but children and mothers do not have equal access to health care providers, teachers, or day care workers who may be capable of making the correct diagnoses. Furthermore, a mother may not be able to report reasonably accurately what she has been told. Unequal access to services or access to lower quality services will lead to biased ascertainment of outcomes. For example, in the 1980s, Newacheck, Budetti, and McManus (1984; Newacheck, Budetti, and Halfon, 1986) examined the doubling since 1960 of the proportion of children reported to have chronic conditions on the National Health Interview Survey. They concluded that much of this doubling did not reflect major shifts in the population, such as deinstitutionalization. Rather, it reflected changes in attitudes and perhaps diagnostic practices among clinicians, and it may have included better ascertainment resulting from better access to medical care for disenfranchised populations through programs like Medicaid. Similarly, diagnoses of learning difficulties may differ depending on the availability of services in the school, teacher demand, and because of changes in the conditions for which special education services may be received. Furthermore, much of the observed increase in diagnoses of autism reflects changes in the special education categories and more awareness of the spectrum (Croen, Grether, Hoogstrate, and Selvin, 2002).

In addition, access to and the quality of services may actually affect outcomes. Interinstitutional variation in the quality of medical and day care and its effect on outcomes are now well established. For example, the NICHD Study of Child Care and Youth Development documented differences in outcomes with higher or lower quality of day care, while Vohr et al. (2004) documented an almost fivefold difference in the rates of cerebral palsy and mental retardation among infants born at 1,000 grams across 12 neonatal intensive care units. The NCS research plan proposes to pay some attention to child care, schools, and religious institutions in the revised section on child health and development, but it is not clear that attention to such issues will also inform other outcomes (e.g., schools as a source of environmental exposures for asthma and as a factor in diets leading to obesity). Clearly, access to early intervention services might alter developmental trajectories, but such services do not seem to be listed for investigation.

Only for asthma does there appear to be a concern about studying the effect of access to medical care. Yet there are many conditions for which variations in developmental screening and other procedures will lead to differential diagnoses and referral for services. For some conditions, like autism, earlier intervention has clearly led to improved outcomes. The NCS does not propose to investigate such factors, despite the well-documented substantial variation that occurs in the services children receive (Mangione-Smith et al., 2007).

Recommendation

Recommendation 3-17: The NCS should add measures of access to and quality of services, including medical care, education, child care, and services, as potential mediators of health and development outcomes and to improve the assessment of information obtained through maternal reports.

Policy Environment

The NCS research plan pays little attention to the implications of policy analysis for the study design. A strong analytic design for policy analysis is the use of "natural experiments," which take advantage of variations in policies across states or communities and across time. An example is the Currie and Gruber (1996a, 1996b) analysis of the impacts of the state expansions of the Medicaid program on use of medical care and birth outcomes. Because states expanded their Medicaid programs in different ways and at different times, the analysis was able to compare outcomes before and after the expansions in each state and control for time effects with the variation in the timing of state implementation. Since interesting policy variation will arise in states, counties, cities, and even neighborhoods, analysts conducting these kinds of policy analyses will need to be able to match detailed information on policies to the geographic location of respondents.

Additional Exposures That Could Be Studied Through Data Linkage

A major potential benefit of the NCS household-based sample design is that a wide range of exposure measures can be obtained by matching existing and future sources of environmental information to respondents' residential addresses. For example, neighborhood demographic and socioeconomic data drawn from the decennial census and often aggregated to the level of the census tract have long been linked in research studies to individuals and households. With the advent of the American Community Survey, which replaces the decennial census long-form sample, demographic and socioeconomic neighborhood information can be linked to individual data more frequently than once a decade (National Research Council, 2007). Crime data from the FBI's Uniform Crime Reporting system can be used to characterize community crime. Environmental data from the Environmental Protection Agency monitoring stations can provide a range of useful measures of environmental pollutants. Data on the policy environment for safety net programs can be derived, for example, from state- or county-specific rules for such programs as Temporary Assistance to Needy

Families. Data on weather conditions, commercial activity, and access to public transportation may all prove useful in characterizing maternal and child exposures. An enormous advantage of these kinds of measures of neighborhood exposures is that they can be added to the NCS database without increasing respondent burden, provided they can be linked to sample members' residential addresses and, for some exposures, dates of interview.

While the research plan acknowledges these kinds of linkages, it does not provide a thoughtful discussion of the steps that need to be taken to optimize NCS linkages to other sources of environmental information. A first priority is to geocode (that is, characterize with standardized measurements of location, such as census block, census tract, city, county, state) all of the residential addresses in which sample children reside over the course of the study. Then a wealth of information stored in other geocoded databases will potentially be available to link to the NCS data. How much residentially linked information the study intends to gather itself and under what conditions geocoded data will be available to analysts to perform their own linkages are not clear.

Given the fundamental importance of a full array of exposure data for testing many of the key study hypotheses and for testing new hypotheses that arise over the course of the study, it is vital that researchers—both inside and outside the designated study centers—be able to access information about all respondents' geographic locations. Such access raises important confidentiality concerns, but these concerns have been met in other national studies, such as the National Longitudinal Survey of Adolescent Health and even Census Bureau data sets, through a variety of mechanisms. One of the most promising for the NCS would appear to be the network of Census Research Data Centers that have been established for researcher use of various census and other governmental sources of information (see Chapter 5).

Turning from the matching tasks to the environmental data themselves, it is important to realize that sources of exposure information are a vital public good for the NCS. The study should encourage researchers, with some combination of internal and adjunct study funding, to compile local exposure data that can be matched to the residential locations of all NCS respondents and to make these data available to all analysts. Westat's data repository would appear to be the logical place in which these geographic data and their documentation would be stored.

Recommendation 3-18: To facilitate linkages of NCS data with environmental exposures from other databases, such as measures of demographics, crime, government programs, and pollution, the NCS should develop a plan for geocoding the residential addresses from prebirth through adulthood of all participating children to

standard census geographic units. In addition, the study should develop arrangements by which researchers, both inside and outside the NCS study centers, can access geocodes for respondent addresses and are encouraged to perform linkages and make their environmental information available to the NCS analysis community. Such arrangements must safeguard the confidentiality of NCS respondents.

4

Study Design, Data Collection, and Analysis

This chapter begins with an overview of the National Children's Study (NCS) design. It then describes, critiques, and makes recommendations on sampling design and data collection plans and their impact on quality control and response burden. Finally, data analysis and dissemination plans developed for the NCS are described and recommendations provided for improvement.

DESIGN SUMMARY

The NCS is designed as a longitudinal investigation of a nationally representative sample of 100,000 births to residents in the United States during the years 2008-2012. The births selected for the NCS will be identified from a probability sample of households chosen with standard survey sampling techniques. A sample cohort of births will be identified from a sample of all noninstitutionalized and cognitively/mentally competent women ages 18-44 who currently live in the national sample of households. The children born to these women during the recruitment period will in turn be followed until they are 21 years old. A variety of physical examinations and interview surveys, conducted in person and by telephone, will be completed during the follow-up period.

The NCS birth cohort will be identified from a stratified cluster sample of households chosen in two or more stages (or selection steps), with random selection used in each stage and the number of stages depending on the pattern of housing in the particular parts of the country that are selected.[1] Some key features of the NCS sample design are presented in Table 4-1. The use of established random selection methods in each sampling stage will ensure that the NCS samples of households, eligible women of childbearing age, and births are national probability samples.

In the first stage, 110 individual counties or small groups of contiguous counties were randomly chosen (by staff at a federal agency partner, Centers for Disease Control and Prevention/National Center for Health Statistics) to serve as the set of primary sampling units (PSUs). Each of these PSUs was selected with probability proportional to size (PPS), with the actual 1999-2002 count of live births serving as the measure of size for each PSU. Since multiple PSUs were chosen in some of the largest metropolitan areas, the 110 PSUs are found in 105 different "study locations or sites." Fieldwork will begin in two waves (in 2009 and 2011) on random subsets of the remaining 98 study locations (the Vanguard Centers will undertake sampling, participant enrollment, and data collection in selected PSUs beginning in mid-2008). Within each sample PSU, 10-15 small groups of neighboring census blocks or block groups, called "segments," are to be randomly chosen. All children born to all eligible female residents of all households located in the 1,100-1,650 sample segments will be included in the study, implying that no subsampling is to be done within sample segments. To improve the efficiency of screening for pregnancies among the women ages 18-44 who are identified in selected households, women who are more likely to become pregnant will be monitored more intensively than the other women. These women may also be recruited through the health care providers they visit during pregnancy and delivery.

Data collection and other survey process activities are to be completed using a common National Institutes of Health organizational structure, in which direct governance is provided by program staff of the National Institute of Child Health and Human Development (NICHD) to a network of centers tied to multiple study locations and a coordinating center, all of which are separately answerable to NICHD. A contract to serve as NCS coordinating center has already been awarded to Westat, Inc., of Rockville, Maryland, as have contracts to 24 centers, including the seven Vanguard Centers. NICHD anticipates the need for about 13 to 15 additional centers.

Developed within this organizational framework, the field operations plan prepared by Westat includes some elements of standardization and

[1]While only two stages of selection are described here, additional stages may be added as needed to accommodate the largest primary sampling units (consisting of more than 500 segments) and high-density urban segments with high-rise residential buildings and other features that make it difficult to achieve targeted household sizes for that stage of selection.

central process control but also the potential for independent local process accountability, since separate contracts would be negotiated by the centers that would conduct the fieldwork in the 105 study locations. Westat's overall responsibility is to integrate its specific process tasks with those done by the other centers. This means that in addition to assuming responsibility for survey instrument development, field staff training, and subsequent sampling activity, Westat must plan and orchestrate all activities surrounding sample recruitment and retention, as well as the collection, processing, and analysis of all data that are obtained during the life of the study. Its coordinating role will require the methods and materials it develops to be incorporated into a written manual of operations. Finally, Westat must also see that all of these activities are completed efficiently, with high quality, and in a timely manner. The general responsibility of individual centers will be to appropriately and effectively apply the methods and use the materials that Westat develops.

SAMPLING DESIGN

NCS sample selection follows well-established principles of population sampling. The most important feature is the strong commitment to random sampling methods in all stages of selection, combined with an effective use of sampling stratification selection with PPS selection in the first sampling stage. This approach is designed to ensure that the initial sample of births will be nationally representative. The extent to which the final data set is representative, of course, will depend on the study team's success in maintaining the statistical integrity of the sample through a prolonged period of follow-up, as noted below. Another critical sampling feature is the use of the same approach (i.e., complete enumeration of a random sample of segments) in each sample PSU. Allowing individual centers to develop their own sampling methods within PSUs could have made analysis of NCS data unnecessarily complex and created the possibility that within-PSU sampling methods were incompatible or inappropriate.

> **Conclusion 4-1:** We strongly endorse the use of probability sampling to select the NCS national sample of births.

Other specific features of the NCS sample design were found to be strategically sound. The expected size of the proposed NCS sample is sufficiently large to justify the equal-probability sample of births, and PPS selection combined with equal numbers of approximately equalized second-stage sampling units (i.e., segments) is the preferred way to yield equal selection probabilities of households, eligible women, and births. While oversampling important population subgroups is often used in national surveys to

TABLE 4-1 Key Features of the Sample Design for the National Children's Study

Stage	**Sampling Unit and Frame Source** (What is being sampled and from what list?)	**Stratification** (Stratify by what? Which sample allocation approach?)
1	**County/small group of contiguous counties** Primary sampling unit (PSU) Defined so that each final PSU is expected to have at least 2,000 births during 1999-2002; subdivided into "geographic units" for subsampling, as needed **Frame:** Grouped listing of 3,141 counties	• 92 roughly equal-sized strata • Stratification by county size and % of births to Hispanic women, % of births to Black women, % of births to Asian women, and % of low-birth-weight newborns
1a (Optional)	**Geographic unit** Formed and subsampled in large PSUs consisting of >500 segments **Frame:** Constructed list of subgroups of segments specially formed with PSU strata	• 18 self-representing (or certainty) PSUs, in 13 of the largest PSUs • Also by geographic proximity somehow (e.g., by sorting on location?)
2	**Segment** Groups of census blocks or block groups (BGs); approximately equal in population size; intermediate selection of "chunks" as needed **Frame:** Constructed list of segments	• 18 self-representing (or certainty) PSUs, in 13 of the largest PSUs • Also "geographically" somehow and (perhaps) by other segment characteristics
2a (Optional)	**Chunk** Formed and subsampled in high-density urban segments with large HH counts **Frame:** Constructed list of chunks in larger segments	• None

NOTES: The eligible-for-study population consists of births to noninstitutionalized and cognitively/mentally competent female residents (ages 18-44) during a designated 4-year enrollment period, based on the woman's location of residence at the time of birth.

Sample Selection (How will random selection be used?)	**Overall Sample Size**
• Some type of without-replacement probability proportional to size (PPS) random selection method — Measure of size = actual 1999-2002 count of the number of births to residents in the county(ies) • Rollout in waves: Vanguard first; then 3 random subsets of the remaining study locations	• 110 PSUs in 105 locations: 7 Vanguard sites + 98 other "study locations" — 18 self-representing PSUs in 13 study locations — One nonself-representing PSU selected by PPS in each of 92 strata
• Select one with PPS in each second-stage stratum — Measure of size = 1999-2002 estimated number of births	• One geographic unit per "stratum"
• Using a field-verified list of USPS carrier addresses, visiting *all* households, enrolling all eligible females and resulting births among those who become pregnant during a 4-year sample recruitment and enrollment period (i.e., no subsampling within segments)	• 10-15 segments selected in most sample PSUs
• One at random within segment • Using a field-verified list of USPS carrier addresses, visiting *all* households, enrolling all births occurring in them during the 4-year enrollment period, as in selected segments (i.e., no subsampling within chunks)	• 1 chunk per segment • 10-15 segments/chunks selected per sample PSU

"Noninstitutionalized" is taken to mean not currently living in an institutional setting (e.g., hospital, nursing home, prison, assisted living facility, etc.), college dormitory, or military barracks. USPS = U.S. Postal Service.

increase the precision of comparisons between subgroups, the likelihood of there being many interesting subgroups to oversample in the NCS (defined by race/ethnicity, exposure to toxic substances or air, etc.), combined with limitations of oversampling methods, made an equal-probability design an appropriate approach for a sample of 100,000 births (Kalsbeek, 2003). The overall size of the proposed sample will support estimates for most important national population subgroups defined by race and ethnicity.

Another positive feature of the proposed design is the use of established geopolitical area units (blocks and block groups) to define segments. Since statistical information from each decennial census is routinely prepared at the block and block group levels, potentially useful ancillary information from the U.S. Census Bureau and other sources will be readily available to describe contextual conditions and differential nonresponse.

A third positive feature is the decision to select the NCS birth cohort sample by screening households for women prior to pregnancy to yield higher sample coverage and avoid the complications of frame multiplicity that result from provider-based approaches (Lesler and Kalsbeek, 1992). While sampling prepregnant women by screening a sample of households may avoid frame problems, it is the more expensive approach of the two because of the relatively low percentage of women screened in a general population sample of households who will actually become pregnant during the sample accrual period. The decision to delineate discovered women by their probability of becoming pregnant by asking them questions whose answers correlate with pregnancy (age, prior birth history, sexual activity, etc.) is conceptually sound, although it was not clear how accurately these data will predict the proportion of monitored women ages 18-44 who will become pregnant. This predicted pregnancy ratio among participating women will be a key component in the estimates of the ratio of the number of households needed to enroll about 1,000 births in each PSU.

> **Conclusion 4-2:** While we endorse the decision to select an equal probability national sample of births as a reasonable strategy given the many key scientific objectives of the NCS, we recognize that a proportionate representation of the study's target population will result in estimates for some subgroups that are not as precise as they would be had those groups been oversampled.

Other features of the NCS sample design raise important issues, however. A first and difficult one is the lower bound of the age criterion for female eligibility. NCS design specifications currently call for monitoring women ages 18-44 in selected households (instead of the more common 15-44 year range to define reproductive age in women's health studies). About 3.4 percent of all births will be omitted. The decision to exclude births to teenage mothers ages 15-17, whose birth outcomes, such as low birth weight (Martin et al., 2007) and infant mortality (Mathews and MacDorman, 2007), are often more unfavorable than for older teens and women in their 20s and 30s may therefore skew findings of health and development outcomes of the children of the youngest women bearing children.

Given the importance of teen pregnancy for both the mothers and their children, it would be desirable to include adolescent mothers in this study, however, the ethical issues in drawing them into the study would present a significant barrier. Although they might be liberated minors for purposes of medical care for themselves and their children, they might not be considered so for purposes of research. The regulations for the protection of human research participants (45 CFR Part 46) present barriers to recruiting this population. Permission of one or both parents of the teen may often be required, whether or not the teen is pregnant. In addition, at least some institutional review boards (IRBs) might conclude that this study poses more than minimal risk given its onerousness, thereby requiring an enhanced level of scrutiny.

Also, these young women may not want to divulge that they are pregnant, and many pregnancies may be unwanted. Interviewing younger women, many of whom live with parents, would also introduce additional ethical issues. Obtaining parental consent to participate in the research might also involve overcoming hurdles such as potentially strained relationships with parents due to the pregnancy. These conditions may adversely affect response rates and retention rates. The panel discussed this issue and, recognizing these difficulties, endorses the decision of the NCS staff to omit births to women ages 15-17 from the NCS.

Another set of concerns with the NCS sample design is related to sampling within PSUs: The current proposal calls for forming approximately equal-sized segments in each PSU and completely enumerating a random sample of 10-15 of them. One of these issues stems from the decision to include all households in selected segments, given the stated goal to enroll approximately 1,000 births per PSU. The difficulty here concerns what is needed to successfully meet the PSU sample size target: either (a) an accurate estimate of the household-to-enrollee ratio for births or (b) the flexibility to vary the number of assigned households in each PSU if accurate estimates of this ratio are unavailable. The problem with completely enumerating selected segments consisting of dozens or hundreds of households is that there is limited flexibility to manipulate the number of assigned sample households to hit the target of 1,000 enrolled births. Of course, one could choose to prematurely stop fieldwork when the sample size target is reached, but this partial implementation of the recruitment protocol for each sample household could seriously compromise the statistical integrity of the NCS birth cohort and thus seriously bias population estimates from a respondent sample that favors the easier-to-recruit cases (Kalsbeek et al., 1994).

Conclusion 4-3: The process of identifying births from a national sample of households is complex and subject to numerous sources of attrition of uncertain magnitude. Because of this, it will be difficult to predict how many households must be initially selected to produce a probability sample of 1,000 births in each of the NCS sites.

Recommendation 4-1: The NCS should consider modifying the sampling design to allow for flexibility in increasing the number of study participants in the event that the estimated number of screened households needed to reach 1,000 births per PSU is incorrect.

One approach would be to choose more segments than currently planned and to subsample from the planned complete enumeration listing of households located in each segment. By subsampling within segments and creating initially assigned and reserve sample components, there would be more flexibility to vary the number of households assigned for fieldwork. Subsampling thus would put less pressure on the accuracy of the household-to-enrolled-birth ratio and the decision as to how large the segments must be. Another complementary strategy to more precisely reach within-PSU sample size targets would be to invest resources during the pilot phase of field operations in the Vanguard Centers on estimating components of the ratio and to observe the determinants of change in this ratio among these sites. Components would correspond to attrition occurring between the following sequential steps of recruitment: assignment, location, screening, and rostering households, followed by female screening, participation in pregnancy monitoring, pregnancy, and giving birth to a live infant.

A related concern is the approach to enumerating households within segments during the "list and screen" phase of sample recruitment. Because of the high cost and less than complete coverage of traditional methods of household listing, the NCS study design calls for an approach that uses vendor-supplied USPS carrier route addresses found in sample segments as a confirmatory source to assist field staff who will be doing within-segment household listing in the traditional way. While this type of adjudicated enumeration of households makes good sense, it should be noted that in our understanding the statistical and practical utility of USPS postal addresses for household listing in area probability samples like the NCS is still in the early stages of development.

Recommendation 4-2: The NCS should consider the proposed household enumeration approach to be experimental and should conduct carefully designed field studies to clearly establish the statistical and practical implications of the proposed adjudicated listing approach.

The investigators should also consider the impact of the dynamic state of the set of residential households during the 4-year time period in which sample enrollment will occur.[2] Specifically, methods should be developed to deal with the differential impact of the relatively more socially dynamic and mobile population subgroups (e.g., young singles, new immigrants) on final sample composition.

Finally, the role of the sample design to ensure a broad diversity of exposure to environmental agents like nonpersistent pesticides, outdoor/indoor air pollution, and aeroallergens was unclear and underdeveloped. While the current design specifications do address *which* particulate exposures are being considered (NCS Research Plan, Vol. 1, p. 6-3 and Tables 6-2 and 6-3), and the discussion of segment stratification within PSUs includes a passing reference to stratifying by "environmental measure" (among others), there is no clear indication of the specific steps that will be taken to ensure that the NCS sample will contain a wide range of environmental chemical exposures of relevance to the study hypotheses. Some examples of sources of ambient environmental source, or exposure, data that can be used to assign segment exposure are the Environmental Protection Agency's (EPA's) Air Quality System (AIS), the Interagency Monitoring of Protected Visual Environments (IMPROVE) network, and point- and area-source primary particulate matter emissions information from the EPA National Emissions Inventory (NET).

Recommendation 4-3: To ensure a diverse exposure profile in the sample, the NCS should consider a careful assessment of variation in ambient exposure to chemical agents within each PSU. If the set of segments in a PSU can be classified by combined exposure to a group of important chemical agents, this information could then be used to form varying exposure-level strata for segment sampling in each PSU and thus ensure a range of ambient exposure to relevant environmental agents.

---

[2]The target population of households thus becomes any household existing for any length of time during the 4-year enrollment period.

DATA COLLECTION

The NCS will rely on approximately 39 study centers, each of which is responsible for data collection in one or more of the 105 study locations. The study centers are selected through a competitive process based on evaluation criteria, such as demonstrated data collection capabilities, the ability to build extensive community networks for recruiting and retaining the sample of women and newborns, and a demonstrated commitment to the protection of individual respondents' information. The study centers will hire and train data collection staff, work to ensure community engagement, and provide scientific support and consultation for the study. During the course of the study, the study centers will collect data at participants' homes and in clinical settings, including the infant's place of delivery. Other data will be collected by mail, telephone, or in person. Biological samples from mother, father, and child as well as air, water, soil, and dust from the child's environment will also be collected.

This is a monumental, multifaceted data collection effort. The multi-organization data collection model relies on government staff directing a central coordinating center that not only provides scientific support for the study's program office but also is the nexus for the management and flow of the myriad data collection activities. The coordinating center must provide the centralized release, control, and maintenance of the sample, as well as the uniform training and quality-control activities to ensure a standardized national data collection. The coordinating center must have strong capabilities in locating sample members as well as in data collection itself, in the event that some study centers fail to meet response rate expectations. The complex decentralized data collection system requires a significant training program for interviewers and their supervisors, frequent communication, monitoring, and follow-up to achieve the goal of creating a standardized and uniform national data set.

While decentralization maximizes the participation and involvement of scientists and researchers in the data collection efforts, it is an unusual model for collecting data from a large national probability population sample. The overarching operational goal of the NCS is to collect completely standardized survey, biological, and environmental data from as many of its 100,000 sampled births as possible. Study sites and centers should be invisible to this process, in the sense that all are contributing completely comparable data to the national collective. A more conventional data collection model, used in such large health-related surveys as the National Longitudinal Study of Adolescent Health, is for a single, well-established data collection entity to control data collection efforts and maximize response rates. Apart from a regional network of supervisors responsible for interviewing, hiring, and quality control, there are no "sites" or "centers" in this model. Of course, elements of the NCS data collection, such as the collection of biological and environmental samples, are extraordinary. But the bulk of the collected data will come from conventional interviews with the parents of the 100,000 children and, eventually, the children themselves. The proposed site-based design threatens the quality of the NCS national data collection efforts if the sites are not managed and monitored very closely by the NCS staff and the coordinating center.

Managing and coordinating the data collection activities of 39 study centers and their quality-control processes will be a formidable challenge. Government staff and the coordinating center will probably develop the detailed specifications for hiring interviewers, for hiring staff to supervise data collection activities, for the uniform training modules that are administered to study center trainers, and for certifying trainers and interviewers to conduct the various data collections and guidelines for monitoring interviewers' workload to lead to uniform and consistent data collections. The coordinating center must ensure the 39 study centers implement quality assurance procedures, including maximizing response rates, for all aspects of the study. This model requires substantial staff involvement to ensure its success, particularly high-level staff with substantial survey research and data collection skills and experience. Also it is likely that there will be site turnover during a 20-year study period from both loss of key personnel and the 5-year competitive renewal process. This turnover will also place a quality/training burden on the coordinating center.

Despite these data quality coordination and control measures, there is likely to be substantial variability across sites in the implementation of the data collection protocols, some of which may threaten NCS study objectives. Despite persistent monitoring of data quality and procedures across all the sites, some sites are likely to fail to meet the high response rate and quality data collection standards required of this study. A data collection model that uses many fewer organizations responsible for data collection is logistically more realistic, practical, and likely to ensure uniform data collection procedures. In particular, using fewer organizations, particularly those that have proven capabilities in collecting high-quality national data in a uniform and consistent manner, would provide a greater likelihood of producing high-quality national data.

Conclusion 4-4: The data collection model adopted by the NCS is complex, will challenge the abilities of the staff and coordinating center to achieve a uniform and consistent national data collection, and may compromise key study objectives.

Data quality will be maximized and missing data minimized if data collection staff receive uniform and comprehensive training along with close supervision and periodic evaluation of their work. The interview situations faced by the data collection staff will be challenging because of the interview length and survey content. Quality-control procedures, such as random repeat interviews or repeat data collection, while burdensome for study participants, are an important component of quality control to identify poor-quality interview staff and incorrectly coded interviews. Study location production and quality-control results, particularly with regard to response rates, are critical to ensuring a uniform data collection system across the nation. The monitoring of productive data-quality goals is critical to a successful data collection, since some study centers may not meet the quality-control guidelines and data collection production goals.

Quality control of the instrument design is equally important, and the use of current best practices is critical, including the careful evaluation of potentially invasive questions, cognitive testing, literacy and cultural sensitivity evaluation, and implementing best practices for translations. The NCS uses several data collection modes for questionnaire data, in addition to collecting biological specimens, environmental samples, and medical examination data. The research plan provides general descriptive information about many aspects of the study, but it does not provide specific information to determine whether best practices are being used in the design and testing of survey instrumentation. In particular, the content of the baseline interview is substantial, but there is little information describing the processes, procedures, and criteria for the items and approaches adopted. While some details may not be appropriate for inclusion in a research plan, an indication that best practices are used in designing and testing questionnaires for the study, including any planned experiments or tests prior to the start of collection, is important. Similarly, little information is available in the research plan about the treatment of language-minority respondents and the procedures for testing questionnaires in languages other than English. Thorough cognitive testing of survey instruments needs to be undertaken at the Vanguard sites for language and cultural validity.

Conclusion 4-5: The NCS research plan does not provide sufficient information on the use of data collection guidelines and quality-control procedures to enable evaluation of the planned implementation of a uniform national data collection system.

To accomplish the goals of the study, a substantial amount of information is needed on a wide variety of topics in household and clinical settings. The estimated response burden through the first 2 years of data collection is approximately 30 hours for women seen prior to pregnancy and 26 hours for women seen in the first trimester.[3] The research plan notes that any single face-to-face data collection will be limited to no more than 4 hours. Other aspects of the collection, such as the clinical setting visits, will require additional time from the study participants. The study requires a substantial commitment of time by each study participant. From a practical point of view, eligible respondents, even after agreeing to participate, will prefer to minimize their participation time. While study staff are aware of the respondent burden associated with the collection and have made efforts to reduce content and consequently interview time, the unique nature of the study and the many stakeholders in the study together will lead to substantial pressure to add additional items to an already ambitious data collection program. The effect of the real or perceived respondent burden can result in lower quality data or no data at all. Staff sensitivity to the public's interest and tolerance of the amount of time the study requires is important to the successful implementation of the study, but there is little information in the research plan to suggest that this issue is a concern.

> **Conclusion 4-6:** The NCS research plan does not address directly the issue of respondent burden, except to say that "some" effort is being made to reduce it, nor does the plan make clear the total number of hours the respondent must commit to the study. In particular, in light of the estimate of the interview length (4 hours) for the baseline interview, a critical collection for the study, the research plan pays little attention to respondent burden and its impact on the quality of the data.

Initial response rates and sample retention are key quality indicators for longitudinal surveys. Initial response rates are particularly important, since so little information is available about the women who fail to participate in the initial wave of data collection. The NCS's target initial response rates for study sites are in the 65-75 percent range. Given the burdensome nature of the data collection, the upper end of this range is ambitious but highly desirable to ensure the overall quality of NSC data. Some of the factors that depress the response rate are the sensitive nature of pregnancy, sexual relations, and fecundity status for some women and the likely disproportionate representation of undocumented migrants and other high-risk groups

[3]This response burden includes the actual home visits, follow-up visits to collect the environmental sampling equipment, time spent completing the self-administered questionnaires, phone calls, and the prenatal ultrasounds. It covers all the contacts from pregnancy (or before) through 24-month phone contact. It also includes the birth visit.

among pregnant women. The response rate goal can be met only if response rates are as high as possible across all sites. Without a more centralized data collection structure or extremely close supervision, study sites are likely to have varied success in implementing the quality-control procedures, obtaining cooperation with the participants, and conducting the interviews.

Recommendation 4-4: The NCS should consider ways in which the survey data collection could be consolidated into a smaller number of highly qualified survey organizations.

While assumptions are based to some extent on the experience of other data collections, incentives to achieve response in other data collections, and the expectation that community involvement will help sample recruitment, the NCS research plan does not explicitly address the best methods and procedures for achieving target baseline response rates. Nor, in any significant way, does it address methods to influence sample units to participate in this long-term study. There is little information on the community engagement model and how the model will be effective for the long-term recruitment of sample participants, nor is there much information on the use of incentives to improve cooperation. Furthermore, there does not appear to be any description of planned experiments on methods to improve recruitment. These issues are critical to ensure a successful baseline interview and need to be addressed.

While attaining a high baseline survey response is critical for the NCS, retaining sample cases over the life of the study is also of central importance. Maintaining the representativeness of the sample over time is key to the acceptance of nationally representative results from the study. Nonparticipation in a longitudinal survey can occur in several ways, and two, in particular, require different approaches to sample retention. First, sample members who initially refuse require interviewer follow-up procedures aimed at overcoming obstacles to their participation. Interviewer training and good supervisory monitoring and controls help in this regard. Second, sample cases who move to locations out of their sample areas must be located and followed to their new residence. Little is said in the research plan about how the study expects to maximize retention of sample cases, particularly regarding follow-up for these two groups of nonrespondents, nor is there any information concerning how long and how often refusal nonresponse and nonlocatable cases are to be included in the sample eligible to be interviewed.

Typically, the largest loss in sample occurs in the early stages of a longitudinal study. The NCS assumption of a 2 percent sample loss each year is a reasonable target after the survey is under way for a few years and provided good locating procedures have been established. The NCS study, however, is likely to experience the most sample loss during the most intensive part of the data collection—the first year—and should devote considerable resources to minimizing these sample losses. While the NCS study staff recognize the sample retention issues and are aware of the need to track sample cases and minimize attrition, there is little discussion of this in the research plan.

> **Conclusion 4-7:** The NCS research plan provides little information concerning best methods for sample recruitment to achieve initial and follow-up target response rates, sample maintenance and sample retention procedures for implementation at the study sites, community involvement plans consistent with the uniform implementation of data collection procedures, or contingency plans to support study sites that do not achieve target response rates.

A successful national data collection program requires the establishment of standards across data collection sites, such as:

- A clear set of measurable production and quality goals for each NCS field site. These measures should include, for example, the following: interviewer training and fieldwork evaluation scores, standard American Association of Public Opinion Research (AAPOR) unit response rates at each sampling level, components of standard AAPOR unit nonresponse rates, profiles of interview length for individual survey instruments, profile of total respondent burden, cutoff rate for completed interviews, item nonresponse rates for key study outcome variables, the annual cohort retention rate, the number of adverse events (e.g., cases of child neglect), as well as any other appropriate measures of the effectiveness of fieldwork.
- A plan for closely and continuously monitoring data collection quality measures, such as those given above. The plan should include guidelines for remedial interventions when sites do not meet expectations, including retraining of data collection specialists, evaluation of supervisory staff or staffing levels at each site, and, if necessary, replacement of site field staff with field staff from the coordinating center or some other qualified organization with relevant survey or data collection experience. Speed in recognizing and resolving these problems is essential.
- A plan for the systematic collection and annotation of regional environmental measures in each of the sites.

Recommendation 4-5: Because of the complexity of the proposed organizational model for data collection and the difficulty of main-

taining the quality and uniformity of data collection procedures across a large number of study sites, the NCS program office should establish and monitor strict standards for enrollment, retention, and data collection at each of the study sites and be prepared to take immediate corrective action if sites do not meet high-quality standards in data collection.

The research plan does not describe how the Vanguard Centers will be used to serve as test sites to identify operational, procedural, and questionnaire design issues prior to their implementation at other study sites. These centers have the unique opportunity to serve as data collection laboratories for the study. While this has occurred on a limited basis, much more could be accomplished. Formal experimentation in Vanguard Centers of procedures or alternative instruments should be encouraged, although the task will be challenging, since the amount of time available prior to full-scale implementation of the study is limited. Operational procedures need to be defined, tested, and refined prior to going into the field. The Vanguard Centers should be the agents for such testing and refining.

For example, a plan for sample size reestimation based on the first year's experience of the Vanguard Centers could be developed. That is, using the data from the first year of enrollment, the study team should reestimate the number of households that must be sampled to attain the overall study goal of 100,000 births. The sample size reestimation will involve quantifying components of attrition in the household sample associated with household screening, recruitment of eligible female residents, their becoming pregnant and giving birth, and enrollment of the birth in the study. The Vanguard Centers can also help develop an initial assessment of the measures that are potentially useful predictors of nonresponse or attrition and whether environmental measures have sufficient variability to support analyses with adequate power.

Recommendation 4-6: The NCS should prepare a plan for monitoring progress of the study in reaching its sample size goals. As part of the plan, the NCS should take advantage of the experience of the Vanguard Centers to evaluate initial enrollment rates, the effectiveness and potential respondent burden of the interview instrument, and the ability of the Vanguard Centers to obtain the required household environmental measures reliably.

A program as multifaceted and complicated as the NCS cannot resolve all the methodological issues related to the various types of data collection that will be used over the life of the study. The study is likely to encounter unsolved questions in how best to measure participant lifestyles or attitudes about environmental exposures, or in how to analyze some of the complex genomic and biological data. The research plan did not explicitly describe any substudies, such as randomizing participants to different incentive treatments, randomizing the order of the formulations of some domains in the interview, or randomizing different strategies for asking sensitive questions.

Determining best practices during the study requires an ongoing program of research and experimentation. Research must be directed to short- and long-term goals, resulting in information on which to base data collection decisions. The conduct of the research must be a priority for the NCS staff and have direct application to the study. While the research plan provides substantial discussion of many topics, there appears to be no formal funded program of methodological research, including pilot studies, that has staff resources assigned to it.

Conclusion 4-8: The NCS research plan does not address the ongoing methodological needs of the study—to study data collection procedures and instruments, conduct experiments, and evaluate the quality of the survey operations and the quality of the data—nor does the plan address the best use of the Vanguard Centers.

Recommendation 4-7: To resolve issues that arise during data collection, the NCS should set aside sufficient resources to maintain an ongoing program of methods research and field experimentation. Among the issues that might be addressed in this research are the reliability and validity of previously untested survey questions and measurement strategies, the effectiveness of sample retention procedures, predictors of response outcomes associated with sample initial recruitment and subsequent annual retention, error implications of unit nonresponse, adjustment strategies for unit nonresponse, and methods for dealing with item nonresponse.

DATA ANALYSIS AND DISSEMINATION

Since all data will be stored on the coordinating center's servers, all analyses will use centrally prepared analytic data sets. In addition to the usual codebooks, standardized formatting, and technical documentation, the data sets prepared by the coordinating center should include data edited, imputed, and weighted using current best practices. Detailed recommendations about analysis methods are not appropriate at this stage, but the following general principles should be followed:

- Data sets should include guidance on the best methods for analysis. The data sets from this study will be analyzed by many teams, often with different goals for an analysis. The guidance should discuss how methods may vary according to whether the goal of an analysis is the estimation of a relationship between exposure and outcome, an estimate of the exposure of a particular set of environmental factors in subsets of the U.S. population, or the prediction of the number of developmental disorders in subsets of the population.

- Data sets should include documentation on the models used to compensate for unit, wave, and item nonresponse. While it will be impossible to predict all uses of the data, the statistical team creating the data sets should work closely with project investigators when constructing the data sets.
- All data sets should include complete documentation on the methods used to impute, the assumptions about analyses used when developing the imputation procedures, and the potential strengths and weaknesses of these methods.
- To help analysts separate the model-building and model-testing phases of their analyses, it would be helpful if the coordinating center were to use the sample design to develop four independent divisions of the sample and include a variable defining these independent quarter samples in the master NCS data set.

Data analysis plans for the main analyses are not part of the research plan. In particular, while many research hypotheses are described, the research plan does not discuss how, when, or by whom the research will be conducted. The analysis plan for studying birth defects from impaired glucose metabolism, distributed at the September 2007 meeting of the panel, added substantial detail to the general analysis plans outlined in the research plan. Even that plan, however, does not have the specificity needed to prepare data collection instruments that both support analysis plans and minimize respondent burden. In smaller, less comprehensive studies, a detailed analysis plan is the primary tool for clarifying or sharpening study questions and for designing data collection instruments. Analysis plans should contain the following features:

- Detailed information about possible predictors, including the scale or units of measurement, and, when possible, references to questions in interviews or assays being conducted by central laboratories.
- Specification of the methods that will be candidates for mitigating the effect of unit nonresponse, study attrition, and item nonresponse.

- Before final approval, each analysis plan should be reviewed by a group consisting of senior scientists and statisticians from the study team who are not involved with the study being reviewed.
- Analysis plans should include strategies to mitigate the model overfitting that can arise when stepwise model selection is routinely used to identify predictors. These strategies could include a prespecification of what variables should always be included in a data set, the use of shrinkage methods such as Lasso (Tibshirani, 1996), and routine use of either training and validation samples or cross-validated model evaluation.

Data from the NCS will provide unprecedented opportunities for the study team and other investigators to learn more about the prevalence of developmental disabilities, the environmental chemicals to which children are exposed, and the relationships between those disabilities and environmental exposures. Past experience with virtually all national data sets is that the research value of the data is maximized when as many skilled analysts as possible are able to access the data for original and replication analyses, and when the peer-review process judges the quality of the analyses performed. In light of this, the national data set that will be gathered as the result of an enormous expenditure of public funds should be made available as soon as is practical to the general scientific community for detailed analyses. The panel acknowledges the formidable challenges in ensuring that the confidentiality of study participants is protected but notes that a number of data dissemination models, such as the Census Bureau's network of data centers, have successfully balanced the potentially competing goals of data access and respondent protection. The current research plan does not discuss any plans for the dissemination of data to the scientific community.

Recommendation 4-8: The NCS should begin planning for the rapid dissemination of the core study data, subject to respondent protection, to the general research community and for supporting the use of the data after dissemination. The costs of implementing this plan should be estimated and set aside in future NCS budgets. Dissemination includes not only the publication of findings through reports and scientific papers and the production of documented data files for researchers, but also active support in the use of NCS data by the broadest possible range of qualified investigators.

The dissemination plan should include the following elements:

- A timeline for making common elements of the data available to the general research community as soon as they have been cleaned and documented.
- A plan for analytic support for investigators wishing to use the data, including analysis consultation, documentation of models used for developing unit nonresponse adjustments to the sampling weights and imputation of data elements subject to missingness, and recommended models for adjusting for measurement error or nondetectable levels in environmental variables.

5

Ethical Procedures and Community Engagement

The National Children's Study (NCS) presents a wide array of ethical issues and poses many opportunities for involving the community in the research effort. Our analysis of these issues rests on the research plan (NCS Research Plan, Vol. 1, Chs. 11, 12, 14) and the NCS Institutional Review Board (IRB) draft document provided by the National Institute of Child Health and Human Development (NICHD). The authors of these documents have laid out plans for addressing numerous issues, such as informed consent, minimal risk, identifying risks, returning information to participants, community engagement, protection of privacy, and sharing of data. Their plans, however, often lack support and generally fail to acknowledge sufficiently that these issues are often enormously complex and at times contentious. (The Federal Advisory Committee, by contrast, in the minutes of their meeting on June 26-27, 2007, does identify the difficulty of the issues posed.) We think that the study faces serious obstacles in seeking approval of the many institutional review boards (IRBs) that it will encounter.

APPROACH TO REVIEW

In the analysis that follows, we relied heavily on the National Academies report *Ethical Considerations for Research on Housing-Related Health Hazards Involving Children* (National Research Council and Institute of Medicine, 2005). This study concluded that compliance with the Federal Regulations for the Protection of Human Research Participants (45 CFR Part 46), while necessary, may not be sufficient for complex studies.

History teaches that more may be required, at least in the court of public opinion. The adverse publicity, litigation, and ultimate settlement of claims arising from the study of partial lead abatement strategies conducted in Baltimore by investigators at the Kennedy Krieger Institute are illustrative. An appropriate metric for the NCS may be whether the study is acceptable in its entirety to the research participants and to society. Even though each individual step arguably poses no more minimal risk when considered independently, it is important to recognize that this study in the aggregate will be quite burdensome and intrusive for the participants.

CRITERIA FOR GIVING INFORMATION TO PARTICIPANTS

A major issue is what information about their own health and development participants will receive. The research plan recognizes that casting the NCS as a strictly observational study is not desirable and that providing information is important for its own sake as well as useful as a means of increasing retention of participants (NCS Research Plan, Vol. 1, Sec. 12.7). The plan states that participants will be given routine information, such as growth parameters and hematocrits. On a purely mechanical level, the plan needs to clarify whether assays will be conducted pursuant to the requirements of the Clinical Laboratories Improvements Amendments of 2007, since conveying noncompliant laboratory results that may influence participants' health care behavior may actually be illegal.

With regard to assessments of participants' health and development, the plan is to provide to participants with findings that are "clinically relevant and actionable," with the caveat that results of genetic tests will be offered to participants and not automatically conveyed (NCS Research Plan, Vol. 1, Sec. 12.7.1). While laudable in many ways, this plan for sharing information warrants further development. The concept of "clinically relevant and actionable" needs to be defined. The research plan and the IRB draft document provide insufficient detail about how decisions will be made about what to disclose, saying only that these decisions will be made centrally to ensure uniformity throughout the country (IRB draft, Sec. 9.4). One measure of the uncertainty that is still present for this important aspect of the study is the statement that "the use of NCS fetal ultrasounds for identification of fetal defects remains under discussion" (IRB draft, Sec. 3.3).

An additional problem is that the point of the NCS is to identify previously unknown factors that influence the health of children. The critical but unacknowledged question, then, is what parents and children need to be told about emerging research findings. The research plan needs to address this issue and indicate whether the NCS plans to rely on standards that have been previously been articulated by such bodies as the National Bioethics Advisory Commission (1999). The NCS documents show little awareness of the vigorous ongoing debates about the scope of the obligation and the desirability of returning research results. For some, the proposed plan will not provide sufficient information to participants. Parents are often eager to get research results, regardless of their clinical utility or the availability of effective intervention. Many investigators have also argued vigorously that many or all research results should be communicated. Some have urged that respect for persons means that participants are entitled to receive results (Shalowitz and Miller, 2005). Another commentator has suggested that research participants should be allowed to decide at the beginning what results they would like, which could then be provided by a secure website (Kohane et al., 2007). Others, however, worry that participants may be misled if they receive a lot of information with uncertain clinical relevance (Beskow et al., 2001), especially since research results often are not replicated (Ioannidis, 2005; Moonesinghe, Khoury, and Janssens, 2007). The balancing between keeping participants informed and maintaining confidentiality is important, and can be dealt with by asking participants whether they want periodic updates.

Recommendation 5-1: The NCS should define the criteria and the process for deciding what individual clinical and research information, such as environmental assessments, test results, and survey scales, will be given to children and their families.

One entity that might assume this responsibility is the Data Safety Monitoring Board, working with the Federal Ethics Committee. One substantive criterion that may be particularly desirable is requiring that internal or external validation studies be available before individual research results are disclosed to children and their families.

Another issue that has not been addressed is that of "look-back liability," a term used to indicate that people who collect clinical data may be liable for damages if they fail to recontact the patient or subject if the data are later found to be clinically relevant (Hirschorn et al., 1999; Sharpe, 1999). The research plan needs to address what responsibility, if any, the study or the local investigator may have to convey the implications of previously collected information to the mother and child.

In addition to deciding what information to convey, more attention needs to be devoted to what research personnel will do when they detect other risks in the course of the study. The research plan does acknowledge that interviewers may detect evidence of child abuse or neglect that, according to statute, must be reported. (We note that many states have more expansive reporting requirements than the reporting intentions identified in the research plan. For example, some states would require that the interviewer report.) Other proposed responses to identified risks may be judged by IRBs to be inadequate, particularly in a study of this magnitude. Many IRBs, for example, require that researchers put into place a system for referral of respondents who are found to be depressed; no such system is considered here. Similarly, it probably is not enough to offer information about help for domestic violence only to respondents who ask for it, which is the current plan (NCS Research Plan, Vol. 1, Sec. 12.3). Suggested international norms require that resources be identified and provided to interviewees regardless of whether they request them (Department of Gender and Women's Health, Family and Community Health, and World Health Organization, 2001). Responsibility for assisting participants at risk cannot be avoided by having them enter their answers directly into computers so that the interviewers do not know their responses—the information to determine risk will be in the central repository. The research plan also seems to lack a specification of how thoroughly the data repository will be screened to identify participants at risk.

A number of factors could be considered in deciding when intervention is appropriate:

- what interventions the participants expect;
- if intervention will not be forthcoming, whether this is truly understood by the participants and their communities; and
- the likely public response to nonintervention.

The plans for and consequences of detecting and responding to risk need to be addressed in the consent process and with the local communities. The study's IRB application, for example, acknowledges that parents need to know that answers to certain questions may trigger reports to children's services (IRB draft, Sec. 9.3).

PROTECTION AND RELEASE OF INFORMATION

The issue of privacy is enormously important for the NCS. At the data collection stage, clear protocols need to be developed for encryption of information on interviewers' laptops or tablet computers and for what to do in the inevitable event that these instruments are lost or stolen (*New York Times*, 2006). The steps currently planned by the coordinating center to protect data during the data collection phase appear to conform to current best practices, but these steps need to be reviewed periodically as best practices evolve. Instructions to interviewers should be exceptionally clear about how to handle sensitive issues that arise regarding the reporting of pregnancy, infertility, and pregnancy termination.

Much more complicated are privacy issues involving the full-sample data set. As with other large-scale longitudinal data sets, the large number of variables about each NCS participant renders impossible the task of somehow "deidentifying" NCS information, especially given the proliferation of other public data sets that can be used for triangulation. The best one can do is to address whether information is more or less identifiable (Malin, 2005; Malin and Airoldi, 2007).

The inability to protect privacy completely, coupled with the enormous scientific and public health value of NCS analysis, poses serious trade-offs for sharing all-site data both with NCS investigators and outside analysts. The NCS proposes a three-tiered approach to data access:

1. NCS investigators would have access to all data after they are stripped of personal identifiers, using "secure systems with safeguards in place to limit access and protect the data."
2. Non-NCS investigators would have to sign data use agreements before receiving access to data that have been more highly redacted.
3. Public-use files would be "maximally de-identified" and may be subject to further restrictions (IRB draft, Sec. 7).

The panel views this approach as fundamentally flawed. First, science and public health are best served by a competitive process in which a number of analysts initiate new analytic studies and attempt to replicate and extend existing and emerging research studies. Second, the enormous expenditure of public funding argues that the NCS data should be considered a public good and available, under strict data protection protocols, to all qualified analysts. Third, given the enormous value of linking environmental and policy information to respondent residential locations, redacted or deidentified data would be worthless for addressing many of the NCS's key hypotheses. Fourth, redacted or deidentified data would hinder or render impossible many valuable replication studies. Fifth, the norm for investigator-driven, large-scale national data collections, many of which are supported at least in part by the National Institutes of Health,[1] is for data release, with appropriate privacy safeguards, as soon as the data are cleaned and documented.

Despite these arguments, providing greater access to NCS investigators may seem attractive as recompense for their "sweat equity" in collecting the data. While recognizing the enormous value of the data collection efforts of NCS investigators, the panel views the compensation argument as flawed. Because data collection sites were selected with probability sampling methods by the NCS to maximize diversity and generalizability (NCS Research

[1]Examples include the National Longitudinal Survey of Adolescent Health, the Health and Retirement Survey, the Fragile Families and Child Wellbeing study, and the Panel Study of Income Dynamics.

Plan, Vol. 1, Secs. 6.1-6.3.6), investigators who were not close to sample clusters had less opportunity to compete to become NCS study centers. Moreover, the study centers competed on their ability to collect the data, not on their ability to analyze it. Finally, the process of collecting data will give sites a significant advantage in carrying out timely analyses of them.

The conditions under which the all-site study data should be released to both insider and outsider analysts warrants very serious attention. The sensitivity of the data and the threat of identifying respondents argues for data release under highly restricted conditions. There are several successful models for releasing highly sensitive data to the research community for general analysis. Perhaps the most promising is the system of secure research data centers developed by the U.S. Census Bureau, which enable qualified researchers to analyze highly sensitive data compiled by the Census Bureau, the Internal Revenue Service, and the Social Security Administration.

All of these considerations argue for releasing data to all analysts, both inside and outside the study centers, but with appropriate privacy safeguards, as soon as they are cleaned and documented. The benefits of such a policy for the advancement of both science and public health greatly outweigh the costs.

Recommendation 5-2: NCS and non-NCS investigators should be given equal access to the full NCS data as soon as they are cleaned and documented. To protect respondent confidentiality, all analyses should be performed with the kind of strict safeguards employed by the Census Bureau research data centers.

The coordinating center plans to obtain a Certificate of Confidentiality, which is issued by the National Institutes of Health to "protect identifiable research information from forced disclosure" (Office of Extramural Research, 2007). Questions exist about the legal efficacy of these certificates, including whether such a certificate would protect against a query under the Data Quality Act or other recently enacted laws that permit greater access to data (Singer, 2005). The NCS may want to consider whether other mechanisms, such as seeking to come within the Confidential Information Protection and Statistical Efficiency Act of 2002 with its protections and penalties, would be more effective. It might also be useful to have some central guidance on HIPAA issues for every site.

IRB REVIEW AUTHORITY

The panel is concerned that the NCS is overly optimistic about the willingness of the study center IRBs to enter into collaborative agreements in which they would defer in whole or in part to other IRBs. Although IRBs are permitted under 45 CFR § 46.114 to defer their authority to another IRB, it seems unlikely that many IRBs would do this in a study with as high a profile as the NCS. Indeed, we predict that IRBs will differ substantially in their assessment of this study, which may seriously complicate its conduct.

INFORMED CONSENT

The research plan and IRB draft document clearly show an awareness of the complexities that surround consent in a study of this nature. The broader plan to evaluate various approaches to consent is excellent, as is the staged approach to obtaining consent, more or less on a need-to-know basis. The proposal to do a trial of a video-assisted consent process in which study personnel are present is sound. However, the way the process is described of allowing respondents to continue after they give an incorrect answer in the video consent process raises questions (IRB draft, Section 12.2). The viewer is left with the impression that respondents will be allowed to continue guessing until they get the right answer, at which time "the presentation will explain why the answer is correct. . . ." More likely, what is meant is that a wrong answer will lead the study personnel to probe further to ensure adequate understanding and agreement, but this procedure should be made more explicit. In another domain, not all would concur with the ethical and legal reasons that are presented for allowing pregnant teens to consent for themselves and their children to participate in this research.

COMMUNITY ENGAGEMENT

The NCS proposes to engage the study community for the primary purpose of recruiting and retaining the sample. One approach to community involvement is called community-based participatory research (Israel et al., 2005; Minkler and Wallerstein, 2003). Under this approach, academic researchers and the community form a partnership and share control of all research processes and activities. Given that the study foci and design are already fixed, it is too late to employ this approach.

Community needs assessments for each study center are being proposed to identify the most pressing local issues of concern. The current NCS documents state that results of these needs assessments will be considered for inclusion in the core or local substudies, yet the documents fail to describe just how such results will be incorporated. Nor is mention made of how diverse community needs will affect data collection efforts across sites and at each of the sites, particularly in light of the NCS's goal of collecting standardized data across all of its sites. Collection of new data that are responsive to a community's needs may raise additional concerns about data collection burden and the use of established and validated measures. Ethical concerns arise if promises for inclusion of measures in the data collection are being made to communities but study centers fail to follow through on such promises.

There are a number of approaches—community collaboration, community cooperation—that, while falling short of full community-based participatory research, should be considered and would yield numerous benefits to the study and to the community. Greater community input into the entire research process has many benefits, some of which are noted by the NCS. These include but are not limited to the promotion of trust between study centers and communities through shared knowledge; more effective and efficient implementation of research, including better recruitment of participants, as noted by the NCS; enhancement of both the quantity and quality of data collected; a sharing of control in the interpretation and presentation of research findings, which may benefit communities as they use the data to address some of their own pressing needs; development of culturally appropriate measurement instruments, which may be useful for those measures that still require validation for diverse populations; and, ultimately, a more accurate understanding of a community's circumstances in comparison to other communities. Moreover, for some of the ethical concerns raised thus far in this chapter, such as the kinds of information that should be provided to participants about their own health or about potential services that may be beneficial to them, community input and perspectives should be sought and represented in decision-making bodies.

Communities should therefore be given opportunities for greater input into study-related issues and collaboration with study centers. The notion of community refers to both the geographic communities in which data are being collected and to members of organizations for whom specific child health issues, such as learning disabilities, asthma, or schizophrenia, are of interest.

We think that the NCS should make a greater effort to engage communities in its research planning. There appear to be no materials that have been developed, nor much discussion in the proposal, about how and to what degree the various communities will be engaged. Moreover, there is no mention of how communities will participate in the implementation and interpretation of the data or whether the data will be available to communities for their own use. Communities can help to identify relevant protective and risk factors that may have been overlooked or underemphasized in the current study design. Given that local needs assessments are being proposed and community-specific data might be collected in the sites, communities should be provided with opportunities for and even resources to support data analyses that are specific to local communities. To take this idea a little further, the NCS might provide opportunities for building community research and data utilization capacity.

Other ethical concerns revolve around previous community-based participatory efforts that have uncovered a high level of discomfort in many communities about the collection of such biomarker samples as urine, DNA, blood, hair, and placenta. Although there is little evidence of discriminatory use of genetic information, the public concerns remain. In the future, specific laws or other legal protections may help to allay these concerns.

A high degree of trust is needed to ensure that a community fully understands (1) the necessity of the collection of the samples, (2) the role that the biological samples will play in answering the research questions and whether the samples will be used for multiple purposes, (3) the procedures involved in collecting the biomarkers, and (4) the eventual interpretation of the information.

Recommendation 5-3: The NCS should engage communities in selected study implementation, data analysis, and data interpretation activities that go beyond recruitment. The NCS should consider requiring every study center to formulate a more detailed plan to engage and collaborate with local communities.

CONCLUSION

In sum, the NCS presents challenges in a broad array of ethical, social, and political domains as a result of its burdensomeness, intrusiveness, and raising of potentially controversial topics relating to children and families. A successful study will require ongoing and, at times, difficult conversations and collaborations with research participants and with the communities in which they live. The NCS should pay particular attention to defining processes for deciding which research results should be returned to participants, to revising the plans for investigators' access to research data, and to developing more robust plans for engaging communities involved in the study.

6

Conclusions and Recommendations

During the past several months the panel has met and reviewed the research plan for the National Children's Study (NCS), various working papers of the study, and additional documents provided by the National Institute of Child Health and Human Development. The panel believes that this landmark study offers an unparalleled opportunity to examine the effects of environmental influences on child health and development, as well as to explore the complex interactions between genes and environments. The database derived from the study will be valuable for investigating the hypotheses described in the research plan as well as additional hypotheses that will evolve.

The critique, suggestions, and recommendations offered in the preceding chapters, therefore, are intended to improve the capabilities of the study to carry out the important mandate of the Children's Health Act of 2000. This chapter highlights the panel's key conclusions and recommendations resulting from its review organized by chapter and subject area.

CHAPTER 2: NCS GOALS, CONCEPTUAL FRAMEWORK, AND CORE HYPOTHESES

Goals

Conclusion 2-1: The stated overall and specific goals for the NCS—and the design of the NCS to achieve those goals—largely reflect the stipulations of the Children's Health Act of 2000. In the broadest terms, the NCS goals and design are responsive to the call in the act for a "national longitudinal study of environmental influences (including physical, chemical, biological, and psychosocial) on children's health and development."

Conclusion 2-2: The large, nationally representative, equal probability sample design, together with the inclusion of a large number of outcome and exposure measures over a long time span, are major strengths of the NCS. In particular, the sample design is an appropriate platform for the study, considering resource constraints, the need to represent all population groups and geographic areas, and the difficulty of devising an alternative disproportionate sampling scheme that would not unduly disadvantage some groups and areas that turn out to be of analytical interest.

Conclusion 2-3: In four overarching areas, the NCS design, as represented in the research plan, is not, or may not be, optimal for achieving the goals of the Children's Health Act. These areas are:

- insufficient attention to understanding disparities in child health and development among population groups of children defined by race, ethnicity, language, socioeconomic status, and geographic area, which the act explicitly mandates;
- inadequate conceptualization of important constructs, including health and development, and an overemphasis on disease and impairment relative to health and functionality and on risk factors relative to protective health-promoting factors;
- impaired data collection schedules and types of measures to support evaluation of some of the effects of chronic and intermittent exposures on child health and development; and
- underappreciation of the challenges to obtaining the highest possible quality of data from an observational design, which include the decentralized data collection structure of the study and limitations on the frequency of home and clinic visits and on the collection of medical and other administrative records for study participants.

Recommendation 2-1: The NCS should give priority attention to seeking ways to bolster the ability of the study to contribute to understanding of health disparities among children in different racial, ethnic, and other population groups, including the reestablishment of a working group to oversee this area and the encouragement of appropriate adjunct studies.

Recommendation 2-2: The NCS should seek resources and develop methods to obtain more frequent in-person measures and medical and other administrative records data on study participants.

Conceptual Framework

Recommendation 2-3: The NCS should clearly define the key constructs of child health and development and more fully develop a conceptual framework for understanding child health and development over the course of infancy, childhood, and adolescence.

Using the Vanguard Centers as Pilots

Recommendation 2-4: We strongly urge the NCS to delay enrollment at new sites to make effective use of initial findings from participant enrollment and data collection in the Vanguard Center sites to improve study procedures, as appropriate, and to refine key concepts, hypotheses, and measures of outcomes and exposures. Throughout the life of the study, the NCS should use the Vanguard Centers to pilot test and experiment with data collection methods and instrumentation.

CHAPTER 3: PRIORITY OUTCOME AND EXPOSURE MEASURES

Pregnancy Outcomes

Recommendation 3-1: The NCS should consider replacing research on subclinical maternal hypothyroidism as a factor in adverse pregnancy outcomes with research on the effects of a broader set of maternal physical and mental health conditions, such as maternal depression, maternal perceived stress, and maternal periodontal disease.

Recommendation 3-2: The NCS should develop refined, detailed protocols for investigating all pregnancy outcomes, specifically a detailed protocol for obtaining information on various types of pregnancy loss, before beginning data collection at the Vanguard Centers, given that pregnancy outcomes are among the first outcomes to be examined; many outcomes lack clarity in measurement; and there are important questions regarding the adequacy of statistical power and the planned data collection (for example, the need for prepregnancy measurements of some exposures).

Neurodevelopment and Behavior and Child Health and Development

Recommendation 3-3: The NCS should develop a clearer rationale for the selection of specific neurodevelopment and behavior disorders to be considered in the study and a clearer conceptual basis for the assessment of normal child health and development trajectories and outcomes. Clarity is needed to guide the choice of outcome measures and exposure measures and the frequency and types of contacts (at the home, in clinics) with study participants in order to obtain the best information possible within resource and burden constraints.

Asthma

Recommendation 3-4: The NCS should develop a clearer rationale for its hypotheses about factors that may increase the incidence of asthma. These should focus on prenatal and early life risk factors.

Obesity and Growth

Recommendation 3-5: The NCS should reevaluate its main hypotheses to be addressed in the study of childhood obesity and consider adopting a broader approach that incorporates social and psychological factors as well as biogenetic ones. Such an approach would help the study identify the constellations of key factors and their interrelationships that are important to understand in order to develop the most effective public health measures to reduce childhood obesity.

Injury

Recommendation 3-6: The NCS should consider replacing research on repeated mild traumatic brain injury (rMTBI) with more nuanced research on other injury-related topics, such as environmental factors in childhood injuries and the effects of clinical response to injury (treatment or nontreatment).

Hormonally Active Agents and Reproductive Development

Recommendation 3-7: The NCS should develop refined and detailed protocols for studying reproductive development outcomes, which, as presented in the research plan, often lack clarity in measurement and research design. Outcomes that are measured at birth for which there is little time to refine research protocols require immediate attention. The NCS should use results from the Vanguard Centers, such as estimates of the prevalence of specific reproductive development outcomes, to assist in protocol development, and it should consider the usefulness of substudies of high-exposure population groups.

Demographic and Socioeconomic Measures

Recommendation 3-8: The NCS should add to its well-planned battery of demographic and socioeconomic measures questions on immigrant generation, languages spoken, and, if possible, the legal status of the parents and child.

Chemical Exposure Measures

Recommendation 3-9: The NCS should consider the use of personal air sampling methods for a subsample of participating women and their children for measuring exposure to air pollutants.

Recommendation 3-10: The NCS should incorporate methodology to measure paternal exposure to environmental chemicals (both persistent and nonpersistent). More generally, the NCS should consider collecting for fathers, not only chemical exposures, but also biological samples and interview data on paternal characteristics that may affect children's health and development to the same degree as it collects such information for mothers.

Physical Exposure Measures

Recommendation 3-11: The NCS should provide a clearer rationale for some of the housing and neighborhood conditions it proposes to measure and revisit its data collection plans to ensure that needed measures are obtained at developmental stages when children may be more vulnerable to risk factors. The goal should be a set of measures and data collection plans that are optimal with regard to analytic utility and response burden.

Psychosocial Exposure Measures

Recommendation 3-12: The NCS should reconsider its psychosocial measures to ensure that they will provide high-quality data for outcomes of interest for child health and development. In the face of resource and respondent burden constraints, the NCS should emphasize the quality and analytic utility of information, even if some measures must be dropped in order to substitute other assessments more desirable on various grounds.

Recommendation 3-13: The NCS should dedicate a portion of funds to support research and development of reliable and valid instruments of key psychosocial measures that are practical and economical to administer.

Biological Exposure Measures

Recommendation 3-14: The NCS should review some of the proposed measures of biological exposures, such as maternal glucose metabolism and child cortisol levels, to ensure that the proposed times for data collection are appropriate for capturing the underlying exposure.

Genetic Measures

Recommendation 3-15: The NCS should adopt a clear mechanism by which genetic association studies are internally and, optimally, externally validated before any results are published or released to the media. The NCS should also revise its proposed "established" candidate gene approach to take advantage of the new information emanating from the current wave of genome-wide association studies, with appropriate replication.

Recommendation 3-16: The NCS should consider consolidating its genetics studies in order to reduce costs and to coordinate the best science at the least cost to the project. One approach would be to simply collect the biological samples and properly store them for later genetic analysis when a better selection of polymorphisms and cost-effective genotyping across studies are possible.

Missing Exposures

Recommendation 3-17: The NCS should add measures of access to and quality of services, including medical care, education, child care, and services, as potential mediators of health and development outcomes and to improve the assessment of information obtained through maternal reports.

Data Linkage

Recommendation 3-18: To facilitate linkages of NCS data with environmental exposures from other databases, such as measures of demographics, crime, government programs, and pollution, the NCS should develop a plan for geocoding the residential addresses from prebirth through adulthood of all participating children to standard census geographic units. In addition, the study should develop arrangements by which researchers, both inside and outside the NCS study centers, can access geocodes for respondent addresses and are encouraged to perform linkages and make their environmental information available to the NCS analysis community. Such arrangements must safeguard the confidentiality of NCS respondents.

CHAPTER 4: STUDY DESIGN, DATA COLLECTION, AND ANALYSIS

Sampling Design

Conclusion 4-1: We strongly endorse the use of probability sampling to select the NCS national sample of births.

Conclusion 4-2: While we endorse the decision to select an equal probability national sample of births as a reasonable strategy given the many key scientific objectives of the NCS, we recognize that a proportionate representation of the study's target population will result in estimates for some subgroups that are not as precise as they would be had those groups been oversampled.

Conclusion 4-3: The process of identifying births from a national sample of households is complex and subject to numerous sources of attrition of uncertain magnitude. Because of this, it will be difficult to predict how many households must be initially selected to produce a probability sample of 1,000 births in each of the NCS sites.

Recommendation 4-1: The NCS should consider modifying the sampling design to allow for flexibility in increasing the number of study participants in the event that the estimated number of screened households needed to reach 1,000 births per primary sampling unit (PSU) is incorrect.

Recommendation 4-2: The NCS should consider the proposed household enumeration approach to be experimental and should conduct carefully designed field studies to clearly establish the statistical and practical implications of the proposed adjudicated listing approach.

Recommendation 4-3: To ensure a diverse exposure profile in the sample, the NCS should consider a careful assessment of variation in ambient exposure to chemical agents within each PSU. If the set of segments in a PSU can be classified by combined exposure to a group of important chemical agents, this information could then be used to form varying exposure-level strata for segment sampling in each PSU and thus ensure a range of ambient exposure to relevant environmental agents.

Data Collection

Conclusion 4-4: The data collection model adopted by the NCS is complex, will challenge the abilities of the staff and coordinating center to achieve a uniform and consistent national data collection, and may compromise key study objectives.

Conclusion 4-5: The NCS research plan does not provide sufficient information on the use of data collection guidelines and quality-control procedures to enable evaluation of the planned implementation of a uniform national data collection system.

Conclusion 4-6: The NCS research plan does not address directly the issue of respondent burden, except to say that "some" effort is being made to reduce it, nor does the plan make clear the total number of hours the respondent must commit to the study. In particular, in light of the estimate of the interview length (4 hours) for the baseline interview, a critical collection for the study, the research plan pays little attention to respondent burden and its impact on the quality of the data.

Conclusion 4-7: The NCS research plan provides little information concerning best methods for sample recruitment to achieve initial and follow-up target response rates, sample maintenance and sample retention procedures for implementation at the study sites, community involvement plans consistent with the uniform implementation of data collection procedures, or contingency plans to support study sites that do not achieve target response rates.

Conclusion 4-8: The NCS research plan does not address the ongoing methodological needs of the study—to study data collection procedures and instruments, conduct experiments, and evaluate the quality of the survey operations and the quality of the data—nor does the plan make the best use of the Vanguard Centers.

Recommendation 4-4: The NCS should consider ways in which the survey data collection could be consolidated into a smaller number of highly qualified survey organizations.

Recommendation 4-5: Because of the complexity of the proposed organizational model for data collection and the difficulty of maintaining the quality and uniformity of data collection procedures across a large number of study sites, the NCS program office should establish and monitor strict standards for enrollment, retention, and data collection at each of the study sites and be prepared to take immediate corrective action if sites do not meet high-quality standards in data collection.

Recommendation 4-6: The NCS should prepare a plan for monitoring progress of the study in reaching its sample size goals. As part of the plan, the NCS should take advantage of the experience of the Vanguard Centers to evaluate initial enrollment rates, the effectiveness and potential respondent burden of the interview instrument, and the ability of the Vanguard Centers to obtain the required household environmental measures reliably.

Recommendation 4-7: To resolve issues that arise during data collection, the NCS should set aside sufficient resources to maintain an ongoing program of methods research and field experimentation. Among the issues that might be addressed in this research are the reliability and validity of previously untested survey questions and measurement strategies, the effectiveness of sample retention procedures, predictors of response outcomes associated with sample initial recruitment and subsequent annual retention, error implications of unit nonresponse, adjustment strategies for unit nonresponse, and methods for dealing with item nonresponse.

Data Analysis and Dissemination

Recommendation 4-8: The NCS should begin planning for the rapid dissemination of the core study data, subject to respondent protection, to the general research community and for supporting

the use of the data after dissemination. The costs of implementing this plan should be estimated and set aside in future NCS budgets. Dissemination includes not only the publication of findings through reports and scientific papers and the production of documented data files for researchers, but also active support in the use of NCS data by the broadest possible range of qualified investigators.

CHAPTER 5: ETHICAL PROCEDURES AND COMMUNITY ENGAGEMENT

Criteria for Giving Information to Participants

Recommendation 5-1: The NCS should define the criteria and the process for deciding what individual clinical and research information, such as environmental assessments, test results, and survey scales, will be given to children and their families.

Protection and Release of Information

Recommendation 5-2: NCS and non-NCS investigators should be given equal access to the full NCS data as soon as they are cleaned and documented. To protect respondent confidentiality, all analyses should be performed with the kind of strict safeguards employed by the Census Bureau research data centers.

Community Engagement

Recommendation 5-3: The NCS should engage communities in selected study implementation, data analysis, and data interpretation activities that go beyond recruitment. The NCS should consider requiring every study center to formulate a more detailed plan to engage and collaborate with local communities.

In summary, it is clear from our review that the NCS offers not only enormous potential, but also a large number of conceptual, methodological, and administrative challenges. In addition, funding uncertainties make it difficult to plan beyond the relatively short period for which funds have been appropriated. Like the scientists associated with the study itself, we are eager for it to succeed. We present our conclusions and recommendations in the hope that, as it goes forward, the NCS will achieve its intended objectives and serve as a model of methodological and substantive contributions to important scientific and policy discussions on children's health and development.

References

Ainsworth, M.D., and Bell, S.M. (1970). Attachment, exploration, and separation: Illustrated by the behavior of one-year-olds in a strange situation. *Child Development, 41*(1), 49-67.

Akinbami, L.J. (2006). The state of childhood asthma, United States, 1980-2005. (Advance data from *Vital and Health Statistics,* no. 381.) Hyattsville, MD: National Center for Health Statistics.

Arbes, S.J., Jr., Gergen, P.J., Vaughen, B., and Zeldin, D.C. (2007). Asthma cases attributable to atopy: Results from the Third National Health and Nutrition Examination Survey. *Journal of Allergy and Clinical Immunology, 120*(5), 1138-1145.

Barlow, S.E., and the Expert Committee. (2007). Expert committee recommendations regarding the prevention, assessment, and treatment of child and adolescent overweight and obesity: Summary report. *Pediatrics, 120*(Suppl. 4:S), 164-192.

Benjamini, Y., and Hochberg, Y. (1995). Controlling the false discovery rate: A practical and powerful approach to multiple testing. *Journal of the Royal Statistical Society, Series B: Methodological, 57*, 289-300.

Ben-Shlomo, Y., and Kuh, D. (2002). A life course approach to chronic disease epidemiology: Conceptual models, empirical challenges and interdisciplinary perspectives. *International Journal of Epidemiology, 31*, 285-293.

Beskow, L.M., Burke, W., Merz, J.F., Barr, P.A., Terry, S., Penchaszadeh, V.B., Gostin, L.O., Gwinn, M., and Khoury, M.J. (2001). Informed consent for population-based research involving genetics. *Journal of the American Medical Association, 286*(18), 2315-2321.

Bloom, B., and Cohen, R.A. (2007). Summary health statistics for U.S. children: National Health Interview Survey, 2006. *Vital and Health Statistics*, Series 10(234). Hyattsville, MD: National Center for Health Statistics.

Bloom, M.S., Buck, L., Schisterman, E.F., Liu, A., and Kostyniak, P.J. (2007). Maternal serum PCB concentrations across critical windows of human development. *Environmental Health Perspectives, 115*, 1320-1324.

Cardon, L.R., and Bell J.I. (2001) Association study designs for complex diseases. *Nature Reviews Genetics*, *2*(2), 91-99.

Centers for Disease Control and Prevention. (2007). *Birth defects*. Available: http://www.cdc.gov/ncbddd/bd/default.htm [accessed October 5, 2007].

Cole, T.J., Bellizzi, M.C., Flegal, K.M., and Dietz, W.H. (2000). Establishing a standard definition for child overweight and obesity worldwide: International survey. *British Medical Journal, 320*, 1240-1243.

Colhoun, H.M., McKeigue, P.M., and Davey Smith, G. (2003). Problems of reporting genetic associations with complex outcomes. *Lancet, 361*, 865-872.

Croen, L.A., Grether, J.K., Hoogstrate, J., and Selvin, S. (2002). The changing prevalence of autism in California. *Journal of Autism and Developmental Disorders, 32,* 207-215.

Currie, M.B., and Gruber, J. (1996a). Health insurance eligibility, utilization of medical care, and child health. *Quaterly Journal of Economics, 111*(2), 431-466.

Currie, M.B., and Gruber, J. (1996b). The efficacy and cost of recent expansions of Medicaid eligibility for pregnant women. *Journal of Political Economy, 104*(6), 1263-1296.

Department of Gender and Women's Health, Family and Community Health, and World Health Organization. (2001). *Putting women first: Ethical and safety recommendations for research on domestic violence against women.* Geneva: World Health Organization.

Drotar, D. (Ed.). (1998). *Measuring health-related quality of life in children and adolescents: Implications for research and practice.* Mahwah, NJ: Lawrence Erlbaum Associates.

Eiser, C., and Morse, R. (2001). Quality-of-life measures in chronic disease of childhood. *Health Technology Assessment, 5*(4), 1-157.

English, S.B., and Butte, A.J. (2007) Evaluation and integration of 49 genome-wide experiments and the prediction of previously unknown obesity-related genes. *Bioinformatics, 23*(21), 2910-2917.

Federal Interagency Forum on Child and Family Statistics. (2005). *America's children: Key national indicators of well-being, 2005.* Washington, DC: U.S. Government Printing Office.

Fox, C.S., Heard-Costa, N., Cupples, L.A., Dupuis, J., Vasan, R.S., and Atwood, L.D. (2007). Genome-wide association to body mass index and waist circumference: The Framingham Heart Study 100K project. *BioMed Central Medical Genetics, 8*(Suppl. 1), S18.

Freedman, M.L., Reich, D., Penney, K.L., McDonald, G.J., Mignault, A.A., Patterson, N., Gabriel, S.B., Topol, E.J., Smoller, J.W., Pato, C.N., Pato, M.T., Petryshen, T.L., Kolone, L.N., Lander, E.S., Sklar, P., Henderson, B., Hirschhorn, J.N., and Altshuler, D. (2004). Assessing the impact of population stratification on genetic association studies. *Nature Genetics*, *36*(4), 388-393.

French, S.A., Story, M., and Perry, C.L. (1995). Self-esteem and obesity in children and adolescents: A literature review. *Obesity Research, 3*, 479-490.

Gibb, B.E., Alloy, L.B., Walshaw, P.D., Conner, J.S., Shen, G.H.C., and Villari, A.G. (2006). Predictors of attributional style change in children. *Journal of Abnormal Child Psychology, 34*(30), 425-439.

Haines, J.L., and Pericak-Vance, M.A. (1998). *Approaches to gene mapping in complex diseases.* New York: John Wiley & Sons.

Hawkins, S.S., and Law, C. (2006). A review of risk factors for overweight in preschool children: A policy perspective. *International Journal of Pediatric Obesity, 1*(4), 195-209.

Hayes, M.G., Pluzhnikov, A., Miyake, K., Sun, Y., Ng, M.C., Roe, C.A., Below, J.E., Nicolae, R.I., Konkashbaev, A., Bell, G.I., Cox, N.J., and Hanis, C.L. (2007). Identification of type 2 diabetes genes in Mexican Americans through genome-wide association studies. *Diabetes, 56*(12), 3033-3044.

Hayes, V.E. (1997). Families and children's chronic conditions: Knowledge development and methodological considerations. *Scholarly Inquiry for Nursing Practice, 11,* 259-290.

Herman-Giddens, M.E., Slora, E.J., Wasserman, R.C., Bourdony, C.J., Bhapkar, M.V., Koch, G.G., and Hasemeier, C.M. (1997). Secondary sexual characteristics and menses in young girls seen in office practice: A study from the PROS network. *Pediatrics, 99*(4), 505-512.

Himes, J.H. (2006). Examining the evidence for recent secular changes in the timing of puberty in U.S. children in light of increases in the prevalence of obesity. *Molecular and Cellular Endocrinology, 254-255,* 13-21.

Hirschorn, K., Fletcher, L.D., Goldmilow, L., Howell, R.R., Lebel, R.R., McCabe, E.R., McGinniss, M.J., Milunsky, A., Pelias, M.Z., Pyeritz, R.E., Sujansky, E., Thomson, B.H., and Zinberg, R.E. (1999). Duty to reconnect. *Genetics in Medicine, 1*(4), 171-172.

Hobbs, N., Perrin, J.M., and Ireys, H.T. (1985). *Chronically ill children and their families.* San Francisco: Jossey-Bass.

Institute of Medicine. (2003). *Unequal treatment: Confronting racial and ethnic disparities in health care.* Committee on Understanding and Eliminating Racial and Ethnic Disparities in Health Care, B.D. Smedley, A.Y. Stith, and A.R. Nelson (Eds.). Washington, DC: The National Academies Press.

Institute of Medicine. (2006). *Preterm birth: Causes, consequences, and prevention.* Committee on Understanding Premature Birth and Assuring Healthy Outcomes, R.E. Behrman and A.S. Butler (Eds.). Washington, DC: The National Academies Press.

Institute of Medicine. (2007). *Nutrition standards for foods in schools: Leading the way toward healthier youth.* Committee on Nutrition Standards for Foods in Schools, V.A. Stallings and A.L. Yaktine (Eds.). Washington, DC: The National Academies Press.

Ioannidis, J.P.A. (2005). Why most published research findings are false. *PLoS Medicine, 2*(8), e124.

Ioannidis, J.P.A., Ntzani, E.E., Trikalinos, T.A., and Contopoulos-Ioannidis, D.G. (2001). Replication validity of genetic association studies. *Nature Genetics, 29*, 306-309.

Israel, B., Eng, E., Schulz, A., and Parker, E. (Eds.). (2005). *Methods in community-based participatory research for health.* San Francisco: Jossey-Bass.

Jacobson, J.L., and Jacobson, S.W. (1996). Intellectual impairment in children exposed to polychlorinated biphenyls in utero. *New England Journal of Medicine, 335*, 783-789.

Kalsbeek, W.D. (2003). Sampling minority groups in health surveys. *Statistics in Medicine, 22*, 1527-1549.

Kalsbeek, W.D., Botman, S.L., Massey, J.T., and Lu, P. (1994). Cost-efficiency and the number of allowable call attempts in the National Health Interview Survey. *Journal of Official Statistics, 10*(2), 133-152.

Kohane, I.S., Mandl, K.D., Taylor, P.L., Holm, I.A., Nigrin, D.L.L., and Kunkel, M. (2007). Medicine reestablishing the researcher-patient compact. *Science, 316*(5826), 836-837.

Krebs, N.F., Himes, J.H., Jacobson, D., Nicklas, T.A., Guilday, P., and Styne, D. (2007). Assessment of child and adolescent overweight and obesity. *Pediatrics, 120,* S193-S228.

Landrigan, P., Garg, A., and Droller, D.B. (2003). Assessing the effects of endocrine disruptors in the National Children's Study. *Environmental Health Perspectives, 111*(13), 1678-1682.

Lesler, J.T., and Kalsbeek, W.D. (1992). *Nonsampling errors in surveys.* New York: John Wiley & Sons.

Lohmueller, K.E., Pearce, C.L., Pike, M., Lander, E.S., and Hirschhorn, J.N. (2003). Meta-analysis of genetic association studies supports a contribution of common variants to susceptibility to common disease. *Nature Genetics, 33*(2), 177-182.

Malin, B.A. (2005). An evaluation of the current state of genomic data privacy protection technology and a roadmap for the future. *Journal of the American Medical Informatics Association, 12*(1), 28-34.

Malin, B.A., and Airoldi, E. (2007). Confidentiality preserving audits of electronic medical record access. *Medinfo, 12*(Pt. 1), 320-324.

Mangione-Smith, R., DeCristofaro, A.H., Setodji, C.M., Keesey, J., Klein, D., Adams, J.L., Schuster, M.A., and McGlynn, E.A. (2007). The quality of ambulatory care delivered to children in the United States. *New England Journal of Medicine, 357*, 1515-1523.

Martin, J.A., Hamilton, B.E., Sutton, P.D., Ventura, S.J., Menacker, F., Kirmeyer, S., and Munson, M.S. (2007). Births: Final data for 2005. *National Vital Statistics Reports, 56*(6). Hyattsville, MD: National Center for Health Statistics.

Mathews, T.J., and MacDorman, M.F. (2007). Infant mortality statistics from the 2004 period linked birth/infant death data set. *National Vital Statistics Reports, 55*(14). Hyattsville, MD: National Center for Health Statistics.

Minkler, M., and Wallerstein, N. (Eds.). (2003). *Community-based participatory research for health.* San Francisco: Jossey-Bass.

Molinari, E., Sartori, A., Ceccarelli, A., and Marchi, S. (2002). Psychological and emotional development, intellectual capabilities, and body image in short normal children. *Journal of Endocrinological Investigation, 25,* 321-328.

Moonesinghe, R., Khoury, M.J., and Janssens, A.C.J.W. (2007). Most published research findings are false—But a little replication goes a long way. *PLoS Medicine, 4*(2), e28.

National Bioethics Advisory Commission. (1999). *Research involving human biological materials: Ethical issues and policy guidance,* Vol. 1. Rockville, MD: Author.

National Institute of Child Health and Human Development. (2007). *The National Children's Study research plan.* June 15 and Supplement September 17.

National Research Council. (2007). *Using the American Community Survey: Benefits and challenges.* Panel on the Functionality and Usability of Data from the American Community Survey, C.F. Citro and G. Kalton (Eds.). Washington, DC: The National Academies Press.

National Research Council and Institute of Medicine. (2000). *From neurons to neighborhoods: The science of early childhood development.* Committee on Integrating the Science of Early Childhood Development, J.P. Shonkoff and D.A. Phillips (Eds.). Washington, DC: National Academy Press.

National Research Council and Institute of Medicine. (2004). *Children's health, the nation's wealth.* Committee on Evaluation of Children's Health, Board on Children, Youth, and Families. Washington, DC: The National Academies Press.

National Research Council and Institute of Medicine. (2005). *Ethical considerations for research on housing-related health hazards involving children.* Committee on Ethical Issues in Housing-Related Health Hazard Research Involving Children, Youth, and Families, B. Lo and M.E. O'Connell (Eds.). Washington, DC: The National Academies Press.

New York Times. (2006, July 6). Size of military data theft grows to affect millions of troops.

Newacheck, P.W., Budetti, P.P., and McManus, P. (1984). Trends in childhood disability. *American Journal of Public Health, 74*, 232-236.

Newacheck, P.W., Budetti, P.P., Halfon, N. (1986). Trends in activity-limiting chronic conditions among children. *American Journal of Public Health, 76*, 178-184.

Norman, R.J., Dewailly, D., Legro, R.S., and Hickey, T.E. (2007). Polycystic ovary syndrome. *Lancet, 370,* 685-697.

Office of Extramural Research. (2007). *Certificates of confidentiality kiosk.* Available: http://grants.nih.gov/grants/policy/coc/ [accessed May 2008].

Ogden, C.L., Flegal, K.M., Carroll, M.D., and Johnson, C.L. (2002). Prevalence and trends in overweight among U.S. children and adolescents. *Journal of the American Medical Association, 288*(14), 1728-1732.

Ogden, C.L., Carroll, M.D., Curtin, L.R., McDowell, M.A., Tabak, C.J., and Flegal, K.M. (2006). Prevalence of overweight and obesity in the United States, 1999-2004. *Journal of the American Medical Association, 295*(13), 1549-1555.

Ottawa Charter for Health Promotion. (1986). First International Conference on Health Promotion. Presented on November 21. Available: http://www.who.int/hpr/NPH/docs/ottawa_charter_hp.pdf [accessed June 2008].

Peeters, A., Beckers, S., Verrijken, A., Roevens, P., Peeters, P., Van Gaal, L., and Van Hul, W. (2007). Variants in the FTO gene are associated with common obesity in the Belgian population. *Molecular Genetics and Metabolism, 93*(4), 481-484.

Seymour, H., Roeper, T., and de Villiers, J. (2003a). *The DELV-CT (criterion-referenced version). The diagnostic evaluation of language variation.* San Antonio, TX: The Psychological Corporation.

Seymour, H., Roeper, T., and de Villiers, J. (2003b). *The DELV-SC (screener). Thediagnostic evaluation of language variation.* San Antonio, TX: The Psychological Corporation.

Shalowitz, D.I., and Miller, F.G. (2005). Disclosing individual results of clinical research applications of respect for participants. *Journal of the American Medical Association, 294,* 737-740.

Sharpe, N.F. (1999). The duty to recontact: Benefit and harm. *American Journal of Human Genetics, 65*(4), 1201-1204.

Singer, E. (2005). Access to research data: Reconciling risks and benefits. *Journal of Law and Policy, 14*(1), 85-114.

Skelton, J.A., DeMattia, L., Miller, L., and Olivier, M. (2006). Obesity and its therapy: From genes to community action. *Pediatric Clinics of North America, 53*(4), 777-794.

Stein, R.E., Bauman, L.J., Westbrook, L.E., Coupey, S.M., and Ireys, H.T. (1993). Framework for identifying children who have chronic conditions: The case for a new definition. *Journal of Pediatrics, 122,* 342-347.

Tanner, J.M. (1962). *Growth at adolescence.* Oxford, England: Blackwell Scientific.

Thompson, R.J., and Gustafson, K.E. (1999). *Adaptation to chronic childhood illness.* Washington, DC: American Psychological Association.

Tibshirani, R. (1996). Regression shrinkage and selection via the Lasso. *Journal of the Royal Statistical Society B, 58,* 267-288.

Toledano, M.B., and Nelson, P.D. (2007). Male fertility-related disorders: Cause for concern or a stalking horse. *Archives of Disease in Childhood, 92,* 565-567.

Vohr, B.R., Wright, L.L., Dusick, A.M., Perritt, W.K., Tyson, J.E., Steichen, J.J., Bauer, C.R., Wilson-Costello, D.E., and Mayes, L.C. (2004). Center differences and outcomes of extremely low birth weight infants. *Pediatrics, 113,* 781-789.

World Health Organization. (1946). *Preamble to the Constitution of the World Health Organization as adopted by the International Health Conference.* New York, June. Signed on 22 July 1946 by representatives of 61 states (Official Records of the World Health Organization, no. 2, p.100) and entered into force on 7 April 1946.

World Health Organization. (2001). *International classification of functioning, disability and health (ICF).* Available: http://www.who.int/classifications/icf/en/ [accessed October 5, 2007].

Biographical Sketches of Panel Members

Samuel H. Preston (*Chair*) is Fredrick J. Warren professor of demography at the University of Pennsylvania. His principal research efforts have been directed toward issues of technical demography, measurement of demographic phenomena, and patterns of human mortality. His expertise and competency lie in actuarial mathematics, statistics, vital statistics, aging, epidemiology, demography, and population sociology. He is a member of the National Academy of Sciences, the Institute of Medicine (IOM), and the American Philosophical Society. He is a fellow of the American Academy of Arts and Sciences, the American Association for the Advancement of Science (AAAS), and the American Statistical Association (ASA). He was named laureate of the International Union for the Scientific Study of Population. He has served on several National Academies committees and is currently a member of the National Research Council (NRC) Committee on National Statistics (CNSTAT) and the Committee on the Social Determinants of Adult Health and Mortality. He has a Ph.D. in economics from Princeton University.

Ellen Wright Clayton is Rosalind E. Franklin professor of genetics and health policy and co-director of the Center for Biomedical Ethics and Society, as well as a professor of pediatrics and a professor of law at Vanderbilt University. Her research and teaching include the ethical, legal, and social implications of developments in genetics. She has been an active participant in policy debates advising the National Human Genomic Research Institute as well as numerous other federal and international bodies on an array of topics, ranging from issues in children's health, including newborn screening, to the ethical conduct of research involving human subjects. She has published two books and numerous scholarly articles and chapters in medical journals, interdisciplinary journals, and law journals on the intersection of law, medicine, and public health. In addition, she has collaborated with faculty in the law and medical schools and the College of Arts and Sciences on interdisciplinary research projects. A frequent teacher and public speaker on medical ethics and other issues, she is currently developing the Law Emphasis Program in the Vanderbilt University Medical School. In addition to teaching in Vanderbilt University's law and medical schools, she is a practicing pediatrician at Vanderbilt Medical Center. Clayton is a member of the IOM and previously served on the IOM Committee on Assessing Interactions Among Social, Behavioral, and Genetic Factors in Health and the IOM Board on Health Sciences Policy; she currently chairs the IOM Committee on a Comprehensive Review of the DHHS Office of Family Planning Title X Program. She has a J.D. from Yale University and an M.D. from Harvard University.

Greg Duncan is Edwina S. Tarry professor in the School of Education and Social Policy, Institute for Policy Research, and faculty fellow in the Institute for Poverty Research at Northwestern University. His research interests are primarily in longitudinal survey research that examines the nature and consequences of poverty and welfare dynamics; neighborhood effects on the development of children and adolescents and other issues involving welfare reform, income distribution, and its consequences for children and adults; and intergenerational consequences on children and adolescents of life in a family using welfare. His current research include dynamic aspects of the incidence of poverty among children and other population groups, the heterogeneous mixture of short- and long-term poverty experiences, the apparent vulnerability of a large segment of American society to at least occasional poverty spells, and the concentration of persistent poverty among certain population subgroups, in particular African Americans. He has served on several committees of the National Academies, including the NRC-IOM Committee on Integrating the Science of Early Childhood Development. He was cochair of the NRC-IOM Committee on Evaluation of Children's Health: Measures of Risks, Protective and Promotional Factors for Assessing Child Health in the Community. He has a Ph.D. in economics from the University of Michigan.

David Harrington is professor of biostatistics in the School of Public Health at Harvard University and chair of the Department of Biostatistics and Computational Biology at the Dana-Farber Cancer Institute. He served as principal investigator of the Statistical Center for the Eastern Cooperative Oncology Group from 1990 to 2000. He conducts research on statistical methods for clinical trials and prospective cohort studies in which the time to an event is a primary outcome. He is also involved in collaborative research on cancer as principal investigator of the Statistical Coordinating Center for the Cancer Care Outcomes Research and Surveillance Consortium. This study is a network of sites around the country, conducting a population-based study of access to and outcomes from cancer care, with special focus on disparities in ethnic subgroups and in the elderly. He has a Ph.D. from the University of Maryland.

Russ Hauser is associate professor of environmental and occupational epidemiology with joint appointments in the Department of Environmental Health and the Department of Epidemiology at Harvard University. His research interests are in the field of reproductive and developmental epidemiology, focusing on the impact of environmental and occupational chemicals on fertility and pregnancy. He is currently conducting a study on the effects of chemicals classified as endocrine disruptors on male and female reproductive health endpoints. He is also conducting a prospective cohort study on children in Chapaevsk, Russia, where he is investigating the relationship of exposure to dioxins and dioxin-like compounds with growth and pubertal development and planning to follow these children to adulthood. Other research activities include studying the relationship between maternal exposure to phthalates and fetal growth and placental function. He served on the IOM Committee on Gulf War and Health. He has an M.D. from the Albert Einstein College of Medicine and M.P.H. and Sc.D. degrees from the Harvard School of Public Health.

William Kalsbeek is professor of biostatistics and director of the Survey Research Unit at the University of North Carolina, Chapel Hill. His experience includes statistical research with the Office of Research and Methodology at the National Center for Health Statistics and at the Sampling Research and Design Center at the Research Triangle Institute in North Carolina. His research interests and areas of expertise are in biostatistics, sample design and research, spinal cord injuries, and assessment. He is well known for his work in survey methods. He is a fellow of the ASA and a member of the American Association of Public Opinion Research and the American Public Health Association. He was a member of the NRC's CNSTAT from 1998 to 2004 and has served as chair of the Panel on Measuring Respirator Use in the Workplace and cochair of the Oversight Committee for the Workshop on Survey Automation. He was a member of the Committee on Sampling Methodologies, the Committee to Review the Social Security Administration's Disability Decision Process Research, and the Panel on the National Health Care Survey. He has M.P.H. and Ph.D. degrees in biostatistics from the University of Michigan.

Sharon Lee Reilly Kardia is director of the Public Health Genetic Program and associate professor of epidemiology in the School of Public Health at the University of Michigan. She is also the co-director of the Michigan Center for Genomics and Public Health, and co-director of the Life Sciences Society Program housed in the University of Michigan, School of Public Health. Her main research interests are in the genomic epidemiology of cardiovascular disease and its risk factors. She is interested in gene-environment and gene-gene interactions and in modeling and in complex relationships between genetic variation, environmental variation, and risk of common chronic diseases. Her work also includes using gene expression and proteomic profiles for molecular classification of tumors and survival analysis in lung and ovarian cancers. As part of her center activity, she is also actively working on moving genetics into chronic disease programs in state departments of health. Kardia has served on IOM and NRC committees, most recently on the IOM Committee on Assessing Interactions among Social, Behavioral, and Genetic Factors in Health. She is currently a member of the Committee on Applications of Toxicogenomic Technologies to Predictive Toxicology and the IOM Roundtable on Translating Genomic-based Research on Health. She has a Ph.D. in human genetics from the University of Michigan, was a postdoctoral fellow in the Department of Microbiology and Immunology and continued postdoctoral work in the Department of Human Genetics.

Daniel Kasprzyk is vice president and managing director of surveys and statistics at Mathematica Policy Research, Inc. He is responsible for overseeing the company's statistical staff and the Washington, DC, survey research staff in Mathematica's Survey and Information Services Division. He is project director for statistical consultation projects that assist the Energy Information Administration, the Agency for Healthcare Research and Quality, and the Internal Revenue Service. He has 30 years' experience managing large-scale sample surveys and carrying out methodological research associated with federal survey programs. He has held various positions on the Survey of Income and Program Participation staff at the Census Bureau. Prior to his current position, he was program director of the elementary and secondary sample survey studies program at the National Center for Education Statistics, where he was responsible for the Schools and Staffing Survey system. He served as the U.S. Department of Education's liaison to the NRC Panel on Estimates of Poverty for Small Geographic Areas. He also chaired the ASA Section on Survey Research Methods as well as serving as officer for the Government Statistics and Social Statistics Sections of the ASA and for the Washington Statistical Society, a chapter of the ASA. He served for 20 years on the Federal Committee on Statistical Methodology. He is an elected member of the International Statistical Institute, a fellow of the ASA, and served as vice president of the ASA. He currently serves as an associate editor for the *Journal of Official Statistics*. He has a Ph.D. in mathematical statistics from the George Washington University.

Milton Kotelchuck is professor and chair emeritus of the Maternal and Child Health Department at Boston University School of Public Health and professor of pediatrics and obstetrics/gynecology at Boston University Medical School. He has extensive experience evaluating public health programs to improve birth outcomes and child health status. His research interests include examination of the adequacy and content of prenatal care, racial disparities in birth outcomes, maternal morbidity, immigrant health, child health services, and health data policy. He developed the widely used Adequacy of Prenatal Care Utilization Index. Currently, he is principal investigator of the Pregnancy to Early Life Longitudinal Database project. He serves on numerous national committees to improve perinatal and child health services, including chairman of the Technical Expert Panel on Evaluation of Healthy Start. Previously, he served as director of the Division of Health Statistics and Research and then assistant commissioner for community health services in the Massachusetts Department of Public Health and was a member of the Massachusetts and North Carolina governors' commissions on the reduction of infant mortality. He is the founding and senior editor of the *Maternal and Child Health Journal* and in 2000 was awarded its first national epidemiology award for "advancing knowledge." He has an M.P.H. in maternal and child health and epidemiology and a Ph.D. in personality and developmental psychology from Harvard University.

Marie C. McCormick is Sumner and Esther Feldberg professor of maternal and child health in the Department of Society, Human Development, and Health in the Harvard School of Public Health. She is also professor of pediatrics at Harvard Medical School and senior associate director of the Infant Follow-up Program at Children's Hospital. She was formerly chair of the Department of Maternal and Child Health. Her research involves epidemiological and health services research investigations in areas related to infant mortality and the outcomes of high-risk neonates. Her current research projects include outcomes of infants experiencing neonatal complications like low birth weight and interventions potentially ameliorating adverse outcomes; evaluation of programs designed to improve the health of families and children; and maternal health and prematurity. She has served on several IOM committees, most recently on the Committee on Understanding Premature Birth and Assuring Health Outcomes. She was a member of the Board on Population Health and Public Health Practice. She currently serves as a member of the NRC Committee on Developmental Outcomes and Assessments for Young Children. She is a member of the IOM. She has an M.D. from the Johns Hopkins Medical School and a Sc.D. from the Bloomberg School of Public Health of Johns Hopkins University.

Nora S. Newcombe is professor of psychology and James H. Glackin distinguished faculty fellow at Temple University. She is a nationally recognized expert on cognitive development and directs the Spatial Intelligence and Learning Center, a science of learning center. Her research interests include memory for early childhood, development of spatial cognition, individual differences in spatial ability, cognitive neuroscience related to these interests, and educational applications of these interests and of cognitive research more generally. She is the author of numerous scholarly chapters and articles on aspects of cognitive development and the author or editor of three books. She has served as editor of the *Journal of Experimental Psychology: General* and as associate editor of *Psychological Bulletin*, as well as on numerous editorial boards and grant review panels. She is a fellow of four divisions of the American Psychological Association, the American Psychological Society, and the AAAS. She has been a visiting professor at the University of Pennsylvania and at Princeton University and a Cattell fellow. She is a past president of the Developmental Psychology Division of the American Psychological Association and president-elect of the Cognitive Development Society and of the Eastern Psychological Foundation. She has a Ph.D. in psychology and social relations from Harvard University.

Patricia O'Campo is Alma and Baxter Ricard chair in inner city health and director of the Centre for Research on Inner City Health at St. Michael's Hospital in Toronto, professor at the University of Toronto, and adjunct professor at the Johns Hopkins Bloomberg School of Public Health. As a social epidemiologist she has been conducting research on the social determinants of health and well-being among women and children for over 18 years. She has focused on methods development as part of her research, including application of multilevel modeling to understand residential and workplace contexts on women's and children's health, the application of concept mapping to increase understanding of how residential neighborhoods influence well-being, and the development of monitoring methods for rare health events in small geographic areas. She has conducted a number of survey-based and longitudinal studies in the areas of the social determinants of adult mental health, intimate partner violence and children's well-being as well as clinic- and community-based evaluations of programs concerning smoking cessation, prevention of perinatal transmission of HIV, and prevention of infant mortality. She has been widely recognized for her contributions to the well-being of women and children through the receipt of early and mid-career awards given by national organizations in the United States and serves on several local, federal, and international committees and boards such as the Board of the Wellesley Institute in Toronto, the national Canadian Perinatal Surveillance System committee, and the national Canadian Institute for Health Research grant review panel on population health. She is a member of the NRC-IOM Board on Children, Youth, and Families. She has a Ph.D. from the Johns Hopkins School of Hygiene and Public Health.